Home Brewing

Ted Bruning

Home Brewing

*A guide to making your own
beer, wine and cider*

National Trust

First published in the United Kingdom in 2011 by
National Trust Books
10 Southcombe Street
London W14 0RA
An imprint of Anova Books Company Ltd

ISBN 9781907892035

A CIP catalogue for this book is available from the British Library.

15 14 13 12 11
10 9 8 7 6 5 4 3 2 1

Reproduction by Mission Productions Ltd, Hong Kong.
Printed and bound by Everbest, China

This book can be ordered direct from the publisher at the website
www.anovabooks.com, or try your local bookshop. Also available
at National Trust shops, including www.nationaltrustbooks.co.uk.

CONTENTS

INTRODUCTION

The traditional reason given, whenever home brewers and winemakers were asked why they started up, was almost always price. A pint of my home-brewed ale, they would proudly say, costs thruppence, when it costs one-and-thruppence in a public house.

That mindset is about as dated as the currency. It belongs to the days of rationing and shortages, when wages were tight and simple commodities were hard to come by; when people collected jam jars and string and re-used wrapping paper and envelopes. The recent recession, and increases in alcohol duty, have gone some way towards bringing thrift back into fashion, but, economic ups and downs notwithstanding, these are more expansive times. True, beer in pubs is expensive; but beer in supermarkets is still cheap. As for wine, it is no longer the middle-class luxury it once was; the selection on supermarket shelves is simply baffling. And these days, if you don't have the money to get what you want right away, there is always plastic.

And yet making your own beer and wine still has a strong appeal. It ties into some very strong trends in modern life: a defiant urge to make and do for yourself; a demand for the authentic rather than the synthetic; a fascination with traditional methods and ingredients; a suspicion that most of the things you can buy are sullied in some way.

Of all the home-based crafts that have arisen from these concerns, brewing and winemaking are among the most accessible in terms of space, equipment and ingredients. Your production can become as sophisticated as you wish, but at entry level the requirements are mostly close at hand. Exclusive use of the kitchen, some fairly basic equipment, much of which can be improvised, and a limited amount of specialised kit are all you need.

Oh, and then there are the raw ingredients. Beer, wine, and cider share the same three basic components – a liquid medium, fermentable sugars and aromatics (flavour) – all mixed up together and fermented with yeast. And that's it. Of course, though, it's not really that simple. Grapes, apples and pears are almost the only fruits that come with all three components in one neat package. With other fruits, as well as with root vegetables, blossoms, nettles and so forth, you have to supply the water and sugar, and in the right balance.

To complicate matters further, the variation of components is almost infinite; and sugary liquids are attractive homes not only for friendly yeast cells but also for a host of other, far less friendly micro-organisms. You will need to master the former and defeat the latter to end up with a beer or wine that is actually pleasant to drink; and to accomplish it you will need two key qualities: first, patience, and second, attention to detail. Sugar content and acidity need to be checked and controlled; correct temperatures have to be achieved and maintained; and, in particular, hygiene must be scrupulous.

But these requirements need not be a deterrent. On the contrary, they are the very qualities that attract people to craft activities of all kinds. To be a good cook, or a good cabinet-maker, or a good model aeroplane builder, you need the capacity to enjoy being utterly absorbed in and take pride in what you're doing. The reward for your painstaking care is a world of choice that commercial concerns don't cater for. Wine-lovers can browse among supermarket shelves and come home with grape wines of all countries, colours, and qualities, but they can never experience a strawberry or a cherry wine. Cider-lovers outside very confined regions of England will be hard pushed to find a place where they can buy a traditional strong, still cider. And beer-lovers will need to live near one of a handful of specialist shops to satisfy a fondness for a near-defunct style such as light mild or barley wine.

And this, perhaps, defines the way in which home brewing and winemaking have changed. In times when choice was more limited, most home winemakers used kits purporting to re-create a claret or a hock, while home brewers favoured similar kits marked mild, or bitter, or brown ale. Today's home brewers and winemakers are more experimental. If they want a decent claret, they can get it from Tesco or Sainsbury's. They would rather explore the huge range of possibilities afforded by the orchards and hedgerows about their homes. Beer-drinkers, likewise, are better served in terms of variety now that supermarkets commonly sell wheat beers, Belgian abbey beers, strong pale ales, even porters. For them, perhaps, authenticity is a greater motive than variety, although there is a whole family of dark beers – strong milds, stock ales, imperial stouts and the like – that are almost impossible to buy.

In summary, then, home crafts of all species, including home brewing and winemaking, are not only enjoyable as hobbies; they also give you a measure of control over your world and the quality of your life. In fact this is more than just a pastime – it's a philosophy in action!

FIRST STEPS: EQUIPMENT

*A chef-turned-microbrewer once confessed to me that
running a brewery was much easier than running a
restaurant kitchen, since the range of ingredients and
equipment was so much more limited.*

Nevertheless, brewing and winemaking are subsets of food
science, so the best place to make your headquarters –
until you get that shed converted – is the kitchen.

YOUR WORKSPACE

Even the most basic kitchen already has nearly everything
you need: electric sockets, running water, wipe-down work
surfaces, plus many of the accessories: weighing scales, sharp
knives, chopping board, pots, pans, even a potato masher and
a rolling pin. But you're going to be busily at work in there for
quite long stretches, so you need to plan your timing around
anyone who needs to use the kitchen to cook.

The kitchen, however, won't provide the space you need for
storage. However humbly you start, if you stick with it you
are going to accrue quite a collection of kit, some of it pretty
bulky. And, once you've made your wine or beer, it's going to
need a dedicated place where it can ferment and mature
unmolested, in some cases for several weeks or, in the case of
some wines, months. Temperature – an even one – is critical.
Wine yeasts prefer to work slowly at room temperature or
marginally above: 15–17°C. The optimum for ale yeasts,
which ferment briskly and quickly, is a little higher: 18–22°C.
Lagers on the other hand are best if they are cold-conditioned

for a long time – anything up to three months at 9–12°C. So for your ale and wine you might think of staking a claim to a corner of the airing cupboard; but there are drawbacks. Most airing cupboards are upstairs, and, while it's fine to carry a 5-litre (1-gallon) glass jar of fermenting wine upstairs, trying it with five times that amount of warm, fermenting beer is probably unwise. A dedicated corner of the kitchen is perhaps the best place to stand your ale fermenter, or perhaps the fabled cupboard under the stairs; it doesn't take up that much space. For lager, the garage will probably do. The key thing, though, is to insulate those fermenters really effectively, because fluctuations of temperature will seriously affect the quality of the end result.

Having earmarked your working space, you need to think about the best layout. With home brewing, especially full-mash brewing, health and safety are genuine issues. The liquids used in the early stages of the process need to be heated to between 80°C and 85°C; then you will be boiling your wort (steeped malt and water) to infuse the hops. You don't want to have to move 25 litres (5 gallons) of hot liquid around the kitchen, so devise a layout of vessels accordingly.

Burns and scalds are not the only health and safety matter involved in brewing: heavy lifting is a consideration too. Lifting can be minimised at most stages of the process simply by moving liquid between vessels bit by bit in a bucket. Scalds can also be an issue with winemaking: some fruits and other fermentables will need boiling (in a ordinary saucepan). More common is the danger of cuts and bruises, since processing your raw materials will involve a lot of chopping and bashing.

Having arranged your storage, workspace, and a secluded spot for your fermenters, it's time to consider equipment. Home brewing and winemaking are wonderlands for people who love gadgets. But to start with, all you need is a plastic bucket and a collection of demijohns (wide-necked glass or plastic jars available both in imperial gallon and 5-litre sizes), together with accessories such as bungs, airlocks, siphon tube, indicator paper, weighing scales, a hydrometer and a thermometer.

First things first – where do you get all this stuff? It's 20 years since Boots stopped selling home brewing and winemaking equipment and consumables. That left the field to specialist shops, and their number has greatly increased. But there still may not be one near you. In that case, don't despair: crank up your computer and search: the web is alive with suppliers who run mail order businesses from their shops.

The next step: let us assume that for your first few forays you are going to play safe and work from a **beer or wine kit**. In the case of brewing, this will take the form of a large can of malt and hop extracts blended to produce either a generic mild, bitter, stout or – increasingly popular these days – a version of a well-known brand such as Woodforde's Wherry. Home winemakers can buy almost identical cans of grape concentrate that will – or should – yield an approximation of claret, say, or burgundy. Specialist shops may also stock concentrates derived from other fruits. All you have to do is dilute your concentrate in your bucket with the prescribed quantity of hot water, and then, when the temperature is correct, add the sachet of yeast (provided with the kit) as directed. Once the first fermentation has died down, you simply siphon your wine into a clean demijohn (a gallon, but let's call it 5 litres), seal it with a bored rubber bung in which an airlock has been inserted, and leave it to get on with it. The brewer can also use demijohns, but a plastic pressure barrel is the best item for the job; when ready, the beer can be dispensed straight from the barrel. Winemakers will probably prefer to serve their guests from proper wine bottles, which at this level of sophistication requires no specialist equipment other than corks (see page 84) and a mop and bucket.

To recap, then, if making wine or beer from a kit (or beer from malt extract), you will need:

For beer:
- A 25-litre (5-gallon) plastic bucket, with close-fitting lid and handle
- A plastic pressure barrel

For wine:
- A plastic bucket, with lid and handle.
- A glass demijohn
- A rubber bung and airlock

For both wine and beer:
- A siphon tube
- A cooking thermometer (glass is best)
- A hydrometer (calibrated glass float) for measuring gravity (see pages 41–3)

All this equipment needs to be completely sterile before use. It should first be cleaned using ordinary washing-up liquid and a stiff brush to remove any dirt that could harbour microbial invaders. Then it needs to be sterilised with a 20-minute immersion in either a solution of **sodium metabisulphite powder** from your home brew shop, made up as directed on the packet; or a 5% solution of ordinary household bleach (rinse very thoroughly indeed!), or **Milton Sterilising Fluid** (from the baby products aisle of your supermarket). Always make up a big batch of sterilising solution and keep it handy for wiping down work surfaces, the exteriors of your vessels and taps, and to dip any spoons, knives, or other utensils in.

Campden tablets – sodium or potassium metabisulphite in tablet form – are an essential part of the home brewer's and winemaker's armoury. Ten of them are equal to a level teaspoon of sodium metabisulphite; they kill most bacteria, inhibit the growth of yeast, and eliminate chlorine. A single crushed tablet added to your wort or must (the fermentable bases for beer and wine respectively) will kill any lingering moulds and bacteria; it quickly deteriorates, though, and after a day or so it will have dissipated and you can safely introduce your yeast. Two or three tablets to 5 litres (1 gallon) will deal with most of the problems you are likely to encounter on the way; another at the end of fermentation will halt the process before it has completely finished if you want some residual sweetness in your wine or cider.

The right brushes are also essential. Bottles in particular need to be thoroughly scrubbed to dislodge any particles that might harbour infection, and a proper bottle brush is the only tool that will reach the entire interior surface. A stiff toothbrush is also useful for awkwardly shaped taps, which harbour the kind of sticky residues that are a microbe's paradise.

MALT EXTRACTS AND COPPERS

For the home brewer, there is an intermediate stage after the kit: malt extracts. These come in cans or bags and are made exactly like kits; the advantage is that you can create your own recipes by blending the extracts of different kinds of malt (see page 24). Brewing from malt extract does require one additional piece of kit, a boiler (always called a copper by brewers), in which the wort is boiled with hops. For more on this, see the full-mash brewing equipment section, page 17.

BEYOND KITS

Kits are easy to use, but it probably won't be long before you want to move on to the next level of difficulty – making beer or wine from raw materials. The large additional task is to extract the flavours and sugars from your raw materials, and the processes vary, according to the materials you choose, from the absurdly simple to the moderately complicated.

ADVANCED WINEMAKING AND CIDERMAKING EQUIPMENT

Let's start with wine. The simplest to prepare of all the various fruits and vegetables you might use for winemaking are, by and large, flowers and blossoms, which require no more than soaking or boiling in a common-or-garden saucepan (although elderflowers have to be rubbed vigorously together first to produce a snowstorm of little white florets).

Soft fruits such as blackberries and raspberries should be wrapped in muslin before being gently crushed with a rolling pin. (Elderberries need to be de-stemmed before crushing.) Custom-made muslin bags are available from your home brew shop, but wrapping the fruit in a muslin square (from the same supermarket aisle as the Milton Fluid) is just as effective. Alternatively, you can dispense with muslin altogether and crush your fruit in the fermenting bucket using your potato masher. You will also need muslin for stone fruits such as plums, peaches and damsons, which have to be bruised (rolled hard enough to break the skin), squeezed, and sieved – the kernels must not be broken as they can produce bitter off-flavours that will ruin the wine. A small single-screw fruit press (a simple manual machine with a screw mechanism that crushes the fruit without needing force, available from all home brew shops or websites), is an inexpensive accessory that is ideal for this job.

Most root vegetables only need chopping; harder fruits and vegetables such as quinces, marrows and potatoes need to be grated – a laborious business. However, a scratter-mill (basically a hopper which feeds the produce on to vertical blades turned either by hand crank or electric motor) will speed up any job that requires grating.

From all this you can see that, muslin apart, most fruit and vegetables can be processed using ordinary kitchen implements. The exception is apples. You can make apple wine by chopping the fruit into small pieces and steeping it in hot water, but the extraction is not particularly efficient and you have to add sugar. To make pure-juice cider, you really do need a scratter-mill to pulp the apples and a screw press to wring the juice out of the pomace or pulp. Miniature versions of both items are available from home brew suppliers.

Incidentally, your kitchen may boast electrical gadgets such as a blender or even a juicer, which you might think would speed up the job. However in most cases they are only big enough to process small batches of fruit and may not, in the long run, save you any time at all. Also, the only metals your ingredients should come into contact with are stainless steel or aluminium; others such as copper and zinc will produce taints that might even be poisonous. Also note: when using a hand-operated screw press, it's important to take a few turns at a time rather than wage a heroic struggle to wring all the extract from your fruit in one mighty effort. Turn the crank until you get a good rush of juice; wait until the rush has subsided, then take a few more turns. And be persistent: even if your pulp or pomace seems to be completely dry, it will still give up a little more juice if you give it a breather and then try again.

So, if making wine or cider from scratch, in addition to the basic equipment, you will need:
- Muslin bags or squares.
- A scratter-mill (optional but advisable).
- A screw press.

If processing fruit and vegetables is a relatively simple business – although often messy – full-mash brewing is of a totally different order of complexity and requires equipment that cannot easily be improvised. First, crushed malt (grist) is steeped in hot water (liquor) in a mash tun to dissolve its sugars and extract its flavour components. When the resulting syrup (wort) has been drained off, the surface of the malt is lightly showered with more hot liquor (sparged) to complete the process. Next, the wort is run into a boiler (copper) where it is boiled with hops. It is then strained into a third vessel where it is allowed to cool before being 'pitched' with yeast (which is simply the brewing term for adding yeast to the unfermented wort). Finally, once the fermentation is finished, it is either bottled or decanted (racked) into casks. So, at least four vessels are required, as well as the means of transferring the liquid from one to another; and temperature is critical throughout. Clearly, this calls for some investment. But it needn't be ruinous. The various items are all available at different levels of sophistication (and cost), from the salvaged and adapted to the custom-made and computerised; how much you spend depends on your ambitions and your proficiency as a scrounger, improviser, and fabricator.

The first item is the **mash tun**. This is the vessel in which the malt (which you will almost certainly buy ready-crushed) is infused in hot liquor until the fermentable sugars are all dissolved. This takes an hour and a half to two hours, at an ideal temperature of around 18°C. The vessel ought to be big enough to hold the whole brew, although it is possible to make less wort and add more hot liquor at the next stage. It needs a

17

tap close to the bottom; and it needs to be well enough insulated to maintain the wort at close to the right temperature for the prescribed period. It also needs a filter or a slotted false bottom that sits above the outflow and prevents the malt from clogging the tap when you run off your wort. A good idea is to fix strings or wires to the filter long enough to hang over the edge of the mash tun so that, once the wort is drained, you can lift up the entire filter and its cargo of spent malt in one simple operation, which saves a lot of scooping out.

You can improvise a mash tun out of a plastic bucket fitted with a tap, using blankets or foam mats for insulation and loading the malt into a nylon mesh bag which you can simply lift out when you want to drain the wort. But the simplest type of mash tun you can buy – an ordinary picnic cool-box fitted with a tap and supplied with a wire mesh false bottom – is so cheap and reliable that you might as well invest in one. If you move up to larger brews than can be accommodated in a cool-box, a more sophisticated version with a thermostatically controlled heating element is available. But thermostats and heating elements are not always reliable, nor rapid enough in their response to falling temperatures to be efficient: if you want to increase your brew-length you are probably better off with a bigger plastic tank wrapped in the kind of jacket used to insulate immersion heaters. And of course, the greater the volume of liquor, the more slowly it will lose heat.

After you have run off your wort, you need to get the last little bit of extract out of the spent malt by sparging, or sprinkling it gently and evenly with hot liquor. The sparging liquor needs to be hotter than the mashing liquor – about 25°C – and expensive mash tuns will come with a rotating sparging arm already fitted. But an ordinary watering can of the size used to water house-plants is perfect for the job.

To charge your mash tun, you will need at least 25 litres (5 gallons) of liquor at the right temperature; and this requires a **hot liquor tank**, which will also supply water at the right temperature for sparging. For years I used a salvaged tea-urn; but here you can double up and use the copper in which the wort will be boiled with hops. Home brew shops will sell you a big plastic bin with a tap and two kettle elements or an immersion heater element fitted at the bottom, which is fine. A branded catering boiler such as a Burco will be more expensive but will last longer and should come with its own thermostat. This will help to keep the hop-boil simmering along at the right temperature for the requisite time; and if your copper is doubling as a hot liquor tank you can turn the thermostat down accordingly, which will save you having to take the temperature of the liquor every few minutes.

Like the mash tun, the **copper** needs an inbuilt filter to prevent the hops from blocking the outflow. You can make do with a metal mesh that fits above the outflow, but the best kind is a matrix of slotted copper tubing that plugs into the outflow: the slots are too small for the hop petals to get into, and the bed of hops sitting on the tubing forms an additional filter that should mean brighter beer. If using a catering boiler, you can put your hops in a mesh bag and hang it inside the boiler, but the extraction is better if the hops can move about freely during the boil.

Once the hops have been added, the wort is transferred to the **fermenter**, which is basically just a big plastic bin, somewhat larger than your brew-length (amount of liquid) to allow space for the formation of a head. Here the wort will turn from flavoured syrup into beer. But, before the yeast is pitched, the wort needs to be **cooled**, for anything above 28–29°C would kill the yeast. You could simply leave it to itself; but it takes a surprisingly long time, during which your wort will be vulnerable to infection. Better, then, to find some way of speeding up the cooling process. One way is to stand your fermenter in a sink of cold water. But that is not much quicker, and it also requires the lifting of large amount of near-boiling wort that has to be stirred to equalise the temperature throughout the vessel. Another method is to immerse a long coil of siphon tubing in the wort, attaching a hose to each end, through which cold water can be circulated. Perhaps the simplest way is to float a string bag filled with ordinary sealed picnic ice-blocks, taking the temperature and giving the wort a good stir every few minutes, and keeping some replacement ice-blocks in the freezer to use as necessary. Once the beer has finished fermenting it undergoes its penultimate transfer, into the **pressure barrel**. This is simply a plastic cask fitted with a valve that allows excess CO_2 to be vented. From there, the final transfer is into your glass.

So, the extra equipment you need for full-mash brewing is:
- A mash tun fitted with a sieve.
- A dual-purpose hot liquor tank and copper.
- A small watering can with a finely perforated rose.
- Two or three ordinary plastic buckets.
- A fermenting bin.
- A pressure barrel.

Before we leave the subject of equipment, it's worth discussing the new generation of ultra-compact brewing systems coming out of Germany.

The one I have seen, the Speidel's Braumeister, imported by Vigo Limited of Devon, is certainly a miracle of modern engineering, as you might expect from one of Germany's top manufacturers of stainless steel beverage tanks. Essentially two vessels – an all-in-one brewery and a separate fermenter – it comes in 20-, 50-, and 200-litre (4½-, 11-, and 44-gallon) capacities and pretty much does the whole thing for you. The 20-litre version stands just 60cm (24in) tall, and the 50-litre size is still only 90cm (3ft). All you have to do is fill it with water, set the programme, and switch it on. Once it has reached mashing temperature you fill the inner sleeve with your chosen blend of malt, and when mashing is finished you lift the whole sleeve out. The wort then comes to the boil, at which point you add your hops. The Braumeister also has an integral cooler, and once it's come down to the right temperature you simply run the wort off into the fermenter, pitch it with yeast, and let nature do the rest.

At £1,400 for the smallest model it's quite a serious investment. However, by all accounts it produces good results and is as flexible in what it can brew as any system you can put together yourself; and besides, when the committed home brewer tots up his or her investment over the years, it could quite easily come to more than £1,400.

BEER: THE RIGHT STUFF

Beer is made with malt, hops, yeast and water – but mostly water. In fact, more than 90 per cent water. And many beer-lovers will say that the quality of the water is critical to the character of the beer. So let's start with water.

WATER AND ITS TREATMENT

It's a myth that Burton beers are brewed with water from the Trent and Guinness is brewed with water from the Liffey. In fact, pretty much all beer is brewed with water from boreholes tapping deep into artesian wells, each with its own mineral profile; it is always said that you cannot replicate a particular beer perfectly unless you use the water from its original well.

The minerals in question depend on the rock through which the rainwater has percolated. In districts of impervious granite or slate the water takes up almost no minerals, and is soft. In districts of sedimentary rock the water dissolves and collects minerals as it passes through, and is hard. The two naturally occurring minerals of interest to the brewer are chalk (calcium carbonate, or calcium bicarbonate in dissolved form) and gypsum (calcium sulphate). One of them is good and one bad.

The different minerals don't confer a flavour as such, but rather affect the behaviour of the water during brewing – which contributes indirectly, of course, to the final character of the beer. For what goes on in the mash tun is a lot more complicated than merely dissolving the sugar present in the malt the way you dissolve sugar in your tea. There are a lot of enzymes at work as well, and the minerals can either help or hinder them at every stage in the brewing process, from extraction during the mash, through hop extraction during the boil, to the performance of the yeast during fermentation. The minerals also contribute significantly to the acidity or pH value of the liquor, which should be mildly acid – a pH value of 5.3 is ideal; 5.2 or 5.4 acceptable; 5.1 or 5.5 too extreme.

The baddie is calcium bicarbonate or **chalk**. If left untreated, it breaks down during the boil and deposits limescale, which can damage your heating elements and fur up your pipes. It also increases the acidity of the mash, thus reducing its efficiency; it chemically brings out the harsher characteristics of the hops; and its ions even interfere with fermentation. The goodie is calcium sulphate or **gypsum**. This isn't broken down by boiling, so it won't damage your brewing equipment. It reduces mash acidity, enhancing extraction. It mellows the astringency of the hops somewhat, allowing you to make the most of their aromatic properties. It even enhances the clarity of the beer. The well water of Burton-upon-Trent is especially rich in gypsum, which is one reason why Burton was always the home of pale ales; and in fact the process of water treatment is almost universally known in the brewing industry as 'Burtonisation'.

You will be brewing with tap water, almost certainly, so, whether you're in a hard or soft water area, you will want to treat your liquor before brewing with it. If you're in a hard water area, give 35–40 litres (8–9 gallons) of water a good boil with a pinch of gypsum for half an hour or so the day before you want to use it. The chalk will settle out during the boil; when it's cool enough to handle safely, simply run the clear water off into the mash tun or any other suitable vessel. Give your boiler a good rinse to get rid of the sediment, and then return the water to the boiler ready for the next day. Also at this stage test the water's acidity with **indicator paper**: if it's above 5.3, add more calcium sulphate in small doses until the required pH value is reached. If you're in a soft water area there is no need for a preliminary boil, as there's no calcium bicarbonate to precipitate. In fact you might even have to add some calcium carbonate (precipitated chalk) to the malt grist before mashing to get the acidity up to the right level.

The last stage in your water treatment programme comes before you start your mash. Bring your water up to the boil again, add 10 grams (2 tsp) of calcium sulphate, and let it boil for half an hour. Then let it cool to the right temperature for mashing, and you're off.

MALT

The malt is the heart of the beer, providing the fermentable materials that create the alcohol, the residual sugars that create the mouthfeel, and many of the flavour components. In Britain we tend to think of malt as deriving entirely from barley. But wheat is now a widely used grain in British brewing; in fact, British wheat beer ranks as a fourth category in its own right alongside Belgian, Bavarian, and Berliner. Typically, the British style doesn't use nearly as much wheat as its continental cousins, because the desired effect is a lighter body rather than the clove, vanilla, banana and bubblegum flavours that come with higher wheat proportions.

Malt is made by soaking barley so that it begins to germinate, thereby starting the process of converting the insoluble starch in the grain into fermentable sugar. It is then dried in a kiln, and the little shoots that have started to sprout are shaken out of it. Before use it has to be ground into a coarse flour or **grist**: traditionally this was the first stage in the brewing process, but home brew shops now supply it ready-milled.

Pale malts of various types, principally lager malt, pale ale malt and wheat malt, kilned at low temperature to preserve the golden colour of the grain, are the base on which nearly all modern beers are founded. The grists of even the blackest of beers generally have pale malts as their main constituents, the colour being supplied by small doses of darker types. Low-temperature kilning preserves the saccharifying enzymes (complex proteins that convert starch into fermentable sugar) in the malt, as well as its colour. This means the malt can be used with other unmalted grains such as wheat, corn and rice.

There is a wonderful variety of malts available to you, listed here from the palest to the darkest.

LAGER MALT: The malt that produces the world's most popular beer style – pale gold, mid-flavoured pilsner derivatives – can also be used as the base malt for other beers, even dark ales. Its ability to convert starch into sugar makes it ideal for use with low-enzyme speciality malts or unmalted grains.

WHEAT MALT: Wheat malt has a high protein content, causing wheat beer's characteristic haze. It also has good head-retention qualities: torrefied (high-roasted) wheat is often used in small quantities for that purpose. It is often used in modest quantities by British brewers to lighten the body of their golden ales. It has to be mashed with grains that supply a husk bed (lager malt is ideal).

PALE ALE MALT: The malt most commonly associated with British ale is kilned at a slightly higher temperature than lager malt for a fuller flavour and darker colour. It also tends to have fewer starch-converting enzymes than lager malt.

RYE MALT: This is used in small quantities for its spicy flavour.

VIENNA MALT: This produces the full-bodied amber-reddish beers that were once popular in Austria but survive mainly in Mexico, Dos Equis and Negra Modelo being noted examples.

MILD ALE MALT: Although most milds are dark, the base malt isn't, creating an amber or copper-coloured wort and producing a dry, full-flavoured beer, with brown, chocolate and/or black malts used sparingly to give the characteristic colour. Mild ale malt is also a good base for stronger ales.

CARAPILS: Kilned at low temperature, this adds sweetness, smoothness and body to pale ales and lagers without affecting the colour or adding caramel notes. Also aids head retention.

MUNICH MALT: An aromatic lager malt that yields a dark reddish-orange wort and a slightly sweet caramel flavour. Munich malt comes in two grades, light and dark.

CRYSTAL MALT: Produced so that most of the starch is not saccharified and is then caramelised during kilning, this is widely used in small additions in ales and lagers to give extra sweet flavours ranging from delicate honey to rich toffee. Also thickens the mouthfeel and creates an attractive bronze colour.

BISCUIT MALT: A base malt toasted and roasted to give the beer a biscuity flavour and a deep amber colour.

AMBER MALT: Similar to mild ale or Vienna, but with more colour and a biscuity flavour. Used in old and brown ales.

BROWN MALT: Traditionally kilned over a wood fire for a smoky flavour, this is used sparingly in stouts and porters.

SPECIAL BELGIAN MALT: A rare malt with a nutty, roasted sweetness that in small quantities enriches brown ales and porters, and in larger proportions adds a plummy, wine-like quality to barley wines and strong winter beers.

CHOCOLATE MALT: Its smooth roasted flavour and brownish-black colour make this irreplaceable in dark ales such as milds, stouts and porters; it can also be used in dark lagers.

BLACK MALT: Even darker, with a sharp, burnt, acidic flavour that can take the sweet edge off some stronger beers.

UNMALTED GRAINS

Not all brewing grains need to be malted. Up to 50% of the grist of a Belgian *witbier* is unmalted wheat, but the malted barley that makes up the rest has enough enzymes to convert the whole mash. Unmalted oats are added to some beers, especially stout, for their husky roughness, which scrubs microscopic particles out of the wort for a silky-smooth finish. Maize and rice are almost flavourless but high in convertible starch. They have no enzymes of their own, but the barley malts they are mashed with have enough for the whole mash. Finally, roasted barley is virtually burnt to an inky black and has an acrid charred-wood flavour, the signature of Irish stouts.

BREWING SUGARS

The use of sugar in brewing is widespread throughout the world. Some brewers oppose the practice because they say sugar is only used as a cheap source of fermentable material to pad out the more expensive malt, and in some cases this is true. American beers in particular use large amounts of rice- or corn-based syrups, and so are characteristically light-bodied.

However, there are other more legitimate and time-honoured uses for various types of sugar in judicious quantities. Corn-derived sugars can be used to boost strength without affecting flavour significantly, or to adjust different brews to achieve consistent strength. A derivative of beet sugar can be used for the same purpose; beet sugar creates an undesirable flavour, but this can be avoided by 'inverting' it – boiling it slowly with water and citric acid to produce a concentrated syrup which is then crystallised. Inverted sugar is much used in Belgium, where it is called *candi*. Corn-derived sugars are also added in small quantities to 'prime' beer before it is bottled or put into cask. Priming sugars prompt a slight secondary fermentation to give the beer a bit of fizz and create a head.

Finally, sugars of various sorts are used for their particular flavour characteristics. A dry beer can be sweetened by the use of non-fermentable lactose. Maltodextrin is a soluble starch that creates a heavier body and richer mouthfeel. Honey is increasingly popular and keeps its distinctive taste throughout the brewing process. And black treacle or molasses might be added to dark beers for its density and depth of flavour.

HOPS

Malt supplies the fermentable base plus some of the richer flavours. It is the oils and resins in the hop, though, that supply the beer's bitterness and aroma as well as the tannin and alpha acids that protect the beer from bacterial infection.

One achievement of small independent brewers in the last 35 years has been to reawaken interest in different hop varieties. Before that, hops had been prized chiefly for their bittering quality, and most of the efforts of researchers and growers had been devoted to developing high-alpha strains and finding new ways of extracting the prized bittering agents. Before the 1980s, hop grists were almost always blended with consistent performance in mind. Today, though, single-hop beers are something you will doubtless want to experiment with.

Hops can be used as whole flowers (cones), dried pellets, or extracts. Bittering or kettle varieties are added early in the boil to maximise extraction of the tannin and acid; aroma or late varieties are added later (late-hopping) for their flavour components. A handful can also be added to the cask (dry-hopping) to confer additional aroma.

The revival of interest in hops means you can now buy varieties from all over the world, especially European lager hops such as Saaz, Hallertau, and Herrsbrucker, and also American varieties such as Willamette, Mount Hood and Cascade, which have been developed to create plenty of aroma but little aftertaste. They are too many to list here; what follows is a summary of the main UK-native types.

ADMIRAL: Admiral has shown itself to be a good replacement for both high-alpha and dual-purpose varieties when used as a kettle hop.

BRAMLING CROSS: Provides fruity blackcurrant and lemon notes and is increasingly popular in speciality beers. If used as a late hop or dry hop, the effect can be very interesting.

CHALLENGER: Gives a fruity, almost scented aroma, with some spicy overtones. A versatile kettle hop for all types of beer, it blends well with other English varieties and is sometimes also used as a late hop and a dry hop.

FIRST GOLD: Has a very attractive lemony aroma but a higher alpha content than traditional aroma hops. Suitable as a general kettle hop and also for late- and dry-hopping in all types of beer. First Gold produces a well-balanced bitterness and a fruity, slightly spicy note.

FUGGLE: The classic English ale variety. Propagated by Richard Fuggle of Brenchley, Kent, in 1875, it became the most widely grown hop in England, until wilt made it almost impossible to grow in Kent and Sussex. It now represents only about nine per cent of the English crop, but is also grown in the USA, mainly in Oregon, and in Slovenia, where is has changed its character and is known as Styrian Golding. Fuggle has a typical English flavour, often blended with Goldings to add roundness and fullness to the palate. It contributes all the essential characteristics of flavour, aroma and balanced bitterness to ales, particularly as its relatively low alpha content means a high hopping rate is needed to achieve desired bitterness levels. Can be used as a distinctive dry hop.

GOLDINGS: Not one variety but a group including Cobbs, Amos's Early Bird, Eastwell Golding, Bramling, Canterbury Golding and Mathons, the strains usually being named after either a grower or the parish where they were first cultivated. Recognised as having the most typical English aroma,

Goldings are also used to late-hop lager, where a delicate aroma is required.

PROGRESS: Similar to the Fuggle but slightly sweeter, and providing a softer bitterness in beers of all types. With its slightly higher alpha content, good value for bitterness if a recipe demands aroma hops for all the bittering elements.

TARGET: Britain's most widely used bittering hop. Provides high yields of alpha acid with a robust aroma. Gives positive floral aromas when used for dry-hopping.

WHITBREAD GOLDING: Provides a distinctive sweet fruit flavour similar to but generally more pronounced than Goldings, and is sometimes used as a distinctive dry hop.

WYE NORTHDOWN: A hop with a very mild, clean flavour, this can be used in all types of beer, with no harshness of palate. Particularly with seedless Northdown, the high level of oil makes this a very distinctive dry hop for full-bodied ales.

The aromatic oils and resins contained in the hop are highly volatile, and, if improperly stored, hops will deteriorate quite quickly. In fact traditional French brewers deliberately let their hops go stale: they only value them for their preservative qualities and prefer to let the flavours of complex blends of malt grists dominate the beer. The British, though, prize the hop's individual character more highly. Here are some tips for getting the most from your hops.

- **Use them fresh**. Buy in small quantities, either fresh or in vacuum-packs, so opened packs don't get stale.
- **Store them in a fridge or even a freezer**. Hop bitterness is lost faster at higher temperatures. Pellets are sometimes easier to store than whole hops.

- **Learn to evaluate hops yourself**. To assess or compare hop aroma, make a hop tea, cool it, and smell.
- **Test-brew with an unfamiliar hop on its own** to see where it fits on the bitterness, aroma and intensity scales. Test it in a 100% pale ale malt beer to show its character.
- **Blend carefully**. Remember, when mixing a grist, that aromas from different hops can sometimes 'average out' and dilute the effect. Single-hop beers usually (but not always) have more aroma definition than mixed-varietal beers.
- **Choose your kettle hop carefully**. Some bittering varieties have strong aromas that may alter the final hop character.
- **Control your brewing process**, or other aromas might hide the hops. Higher alcohols, esters, diacetyl and hydrogen sulphide will all disguise or overlay your hop aroma.

YEAST

Yeast is the marvellous micro-organism that transforms hopped malt syrup into beer, which it does by digesting the sugar and excreting carbon dioxide and alcohol. Mankind has been enjoying alcohol for millennia, knowing for most of that time what yeast did but not how. Medieval ale-wives were adept at cropping and propagating the barm (foamy scum) that formed on the top of their tubs; but they thought it was magical, and rendered the magic safely orthodox by calling the yeast 'Godisgood'. Only in the 19th century did Louis Pasteur finally unravel the mystery: yeast, he discovered, was nothing more magical than a variety of fungus with an especially sweet tooth.

Yeast works in two stages. The first or **aerobic fermentation** is the spectacular one which throws the thick frothy head; at this stage the yeast is not producing alcohol but is digesting sugar in order to multiply, and it needs plenty of oxygen to help in the process. This is why wort has to be aerated. After this phase, the yeast population will have multiplied enough to enter phase two, **anaerobic fermentation**, during which the alcohol is produced and air must be excluded.

Brewer's yeast is generally divided into two categories, ale or top-fermenting yeast and lager or bottom-fermenting yeast, so called because ale yeast tends to rise to the top while lager yeast tends to sink. The terms are not entirely accurate because in both cases there are yeast cells present and active throughout the wort, but they are universally understood. Ale yeast ferments rapidly at a warm temperature, while lager yeast continues to work slowly for up to three months at a much lower temperature. Within these categories there are countless different strains, because all breweries propagate their own yeast in jealous isolation; they even send samples to yeast banks for safekeeping in case their own stock becomes infected and has to be replaced. It is always claimed that all these different strains are partly responsible for the house signature of each brewery's beers, although you need a very sensitive nose and palate to detect it. But if your nose and palate are that sensitive, you might want to develop a strain of your own: it's tricky to keep, but it's a talking point!

Probably, though, you will buy your yeast fresh for each brew from your home brew shop, either dried in individual sachets or as a liquid suspension, to be made up and added as directed. You can also use the yeast sediment found in bottle-conditioned beers, or even visit your local microbrewer and scoop a couple of cupfuls off the top of his fermenting beer – bring a sterilised Thermos flask with you to transport it in.

BEER: BREWING

With all your ingredients assembled, your equipment and working areas thoroughly sterilised and your water duly treated, you're almost ready to 'mash in'.

First, though – and this ought to be done a couple of days beforehand – you need to prepare your yeast starter.

YEAST STARTER

Your yeast really needs to hit the ground running so that it can grow and multiply as soon as it is pitched (added). The sooner the wort can develop a dense, rocky head, the sooner it is protected from any airborne bacteria that might be floating about. And the sooner the yeast can get through the first (aerobic) phase of fermentation, the sooner it can get started on the anaerobic phase during which it produces alcohol.

First, thoroughly sterilise a large jar. A family-sized jam jar, an old-fashioned milk bottle or a baby's feeding bottle will all do nicely. Then boil a piece of open-weave cloth to cover the mouth of your jar or bottle.

Next, boil 225ml (half a pint) of water with about 50 grams (2oz) of sugar or malt extract for a few minutes until you have a thin syrup. Once it has cooled enough to handle safely, pour it into your jar, covering the mouth with foil, and bring it down to room temperature by standing it in cold water. When it has reached room temperature, shake it well to aerate it, then add a sachet of yeast and cover the mouth with your sterile cloth, secured with a rubber band. Leave it in a warm place and give it another careful shake two or three times a day, because during the first phase yeast needs air to multiply rapidly. It should soon begin to show signs of growth, and within two or three days you will have a vigorous culture.

If using the yeast sediment from a bottle conditioned beer, pour the beer very carefully into a glass, leaving a centimetre (half an inch) or so in the bottle; swirl the bottle vigorously to suspend the sediment in the remaining beer; then tip it into your syrup. Yeast procured in this way doesn't always start: the yeast-count might be low, it might be a fairly docile strain, it might be too old, or it might have been damaged by ultraviolet light (which is why the best beer bottles are brown), so have a standby culture in reserve.

MASHING

Your hot liquor tank with its 35–40 litres (8 9 gallons) of pre-treated water (now properly called liquor) should be on the kitchen worktop, with your mash tun (on a towel or cloth – there will be drips!) placed firmly on a sturdy stool, trestle table or camping table directly under the tank's tap. Give the liquor a good long stir to aerate it, using a paddle or even an electric blender, and then bring it up to about 80°C. Now run the amount of hot liquor stipulated by your recipe quickly into the mash tun, which process should aerate it further. Put the lid on the tun and wait a few minutes while the temperature sinks to 72°C.

Gently pour in your grist, stirring it gently but thoroughly to achieve a thick porridge in which there are no dry pockets. Don't overstir, though – if the grist gets waterlogged it will sink to the bottom of the tun in a thick mass that won't ferment properly. Now take the temperature of the wort: it needs to be about 66–68°C, and you can add boiling or cold water to adjust it as necessary. Replace the lid along with any additional insulation, and leave it all to mash for 90 minutes.

While the grist and the liquor are getting acquainted, boost the remaining liquor in the tank up to 85°C for sparging. As soon as you're ready to start running off your wort, fill your watering can with the hot sparging water, and pour off and reserve any remaining hot liquor into a separate bucket.

RUNNING OFF AND SPARGING

Running the wort off into the copper (as your emptied hot liquor tank has now become) is quite a tricky process until you get used to it. You have to empty the mash tun in stages, because to run the wort directly into the copper would mean putting the copper on the kitchen floor under the mash tun's outflow tap and, once it's full of hot wort, somehow hoisting it back up on to the worktop. Clearly, this is to be avoided. Instead, place an ordinary plastic bucket (again, on a towel) on the floor directly under the tap, half-fill it, decant it into the copper (which is still on the worktop); repeat as necessary.

To complicate matters, you have to sparge your grains – sprinkling them gently with hot water (85°C), using your watering can, to flush out the last lurking sugars – at the same time and at roughly at the same rate as the wort is draining. This means only half-opening the tap and monitoring the outflow even as you wave your can from side to side over the mash tun to make sure the whole surface of the wort gets an even sprinkling. The sprinkling must be done gently, because the grist will act as a filter, trapping fine particles and ensuring a nice bright beer, so running off is of necessity a slow process. If you find sparging too fiddly, a second-best method is to transfer the wort via the first bucket into the copper as above, turn off the mash tun's tap (!) and pour a couple of litres of the remaining hot liquor from your reserve bucket over the grist. Cover the tun and leave it to stand for 15–20 minutes, then run it off into the bucket and add it to the copper.

A good tip – again, to increase aeration – is to pour the wort from the bucket into the copper from as much of a height as you can safely (and accurately) manage.

THE BOIL

Once the wort has been transferred safely to the copper, top it up to the desired level with the rest of your reserved hot liquor, remembering to leave enough head-room for the foam that will be produced during the 90-minute boil.

As soon as the wort comes to the boil, add your kettle (or bittering or early) hops. Sugars and syrups should be introduced as per the recipe, and then, about 10 minutes before the end of the boil, add your aromatic hops. This is also the time to add Irish moss, if you're using it. (This is not a flavouring but a fining agent (see page 40), precipitating any haze-forming proteins and making a pale beer bright and shiny.) Always remember, though, that you're dealing with a large quantity of boiling water, so remove and replace the lid of the copper with extreme care and add hops and sugars gently and without splashing. No need to stir.

After your 90-minute boil, let the copper cool to 80°C or thereabouts and stir in any further hops your recipe requires.

COOLING AND FERMENTATION

Your fermenter should now be installed on the sturdy stool or table previously occupied by the mash tun. As soon as the wort is cool enough to transfer safely, simply run it into the fermenter. Whole hops or cones will prove their advantage over pellets here: as you drain the copper they form a mat on the copper matrix at the bottom that ought to filter out many of the microscopic particles that would otherwise affect the clarity of the final product.

Now go through your chosen cooling procedure as described in the 'Equipment' chapter (see page 20), stirring the wort well before using the thermometer to equalise the temperature throughout the vessel. Once the temperature is down to 28–29°C it is almost ready for the yeast to be added. Before you do that, though, assess its alcoholic strength using your hydrometer (see pages 41–3) and make any adjustment with a little cold water. Next, it needs to be thoroughly aerated. You can do this by running a few litres into one of your buckets, swirling it as vigorously as possible a few times, and then tipping it back into the fermenter from a height. Do this two or three times and you should have introduced enough air for the yeast to work on, so you can add it now.

Once you've added your yeast, you're going to come across the one bit of heavy lifting that cannot be avoided. You really can't leave the fermenter on a stool in the middle of the kitchen floor for the week or so it will take before it's ready to rack into cask. You need to be able to move it to wherever you've set aside, so make sure you've bought a vessel with a good strong handle. Some kind of trolley might be useful too.

To protect the beer from infection while the head develops, keep the lid on the fermenter. Once the head forms, carefully scoop out any dark flecks and bits of hop; you don't actually need to replace the lid, but it's probably wise: if the fermentation is taking place in the cupboard under the stairs there's always the risk of bits of plaster and dead spiders falling into it; if it's in the kitchen, there will be tiny airborne droplets of grease and other cooking by-products floating about which you don't really want in your beer. If the head is too vigorous and threatens to overspill the top of the fermenter, or if it's not vigorous enough and looks likely to collapse into the beer, by all means skim it. Otherwise it's not necessary and you probably ought to leave well alone.

NOTE: STUCK FERMENTATION
During the first stage of fermentation it is important to keep aerating the wort to prevent it from 'sticking': give it a vigorous stir every day for the first three or four days. Take a daily hydrometer reading, and if the gravity stops falling the fermentation is probably stuck. In which case rouse the yeast with another good stir; and keep spare sachets of yeast handy.

RACKING, FINING AND MATURATION

Test the gravity of the beer using your hydrometer every day. When it is either at or near the final gravity stated in the recipe and has been stable for a day, it's time to **rack** the beer into your pressure barrel. Some people rack their beer using a siphon, but siphons are a messy and fiddly alternative to the more obvious method of simply returning your fermenter to the kitchen worktop, standing your collecting vessel on the trusty old sturdy stool or table, and using the tap. Siphoning also creates the risk of air coming into contact with the beer and of sediment creeping into the barrel.

You can either rack your fermented beer directly into the barrel or, preferably, into an interim vessel to allow any excess yeast to settle out for a day or two. This makes a little extra work, but it does lead to a more refined result without that bready 'home brew' taste. Alternatively, leave the beer in the fermenter for a couple of days after fermentation has stopped. The barrel should be stored in a cool, dark place and, if it wasn't quite full when you racked your beer into it, it should be vented after two days to allow the air to be displaced by a layer of the protective CO_2 given off by the slight residual fermentation that will be going on.

The beer isn't quite ready to drink yet, though. There are two more steps to go through before you introduce glass to spigot (the tap on the cask): fining and maturation.

Fining is the process of clearing the beer of microscopic particles of yeast in the cask by pouring in an agent such as collagen (isinglass, derived from the swim-bladders of fish) or unflavoured gelatine. Isinglass is positively charged and attracts the negatively charged yeast particles as it slowly settles through the beer. It can be messy to use and, unless it's very fresh, doesn't always work that well. Gelatine is easier to use but less effective. Fining is an unnecessary process if you intend to give the beer a decent maturation period or if you plan to bottle it: it's a short cut, really, used by commercial brewers to speed up the clearing of the beer; and you only need to fine your beer if you want to drink it soon. The fining agent – either a gel or a powder – should be made up with a little of your beer according to the instructions on the packet: 500ml to a litre (1–2 pints) of dissolved finings will treat five gallons of beer. Pour it in gently, stir, then leave it for 12 hours or overnight.

A better option is to allow the surviving yeast plenty of time to munch its way slowly through any remaining carbohydrates in the beer. A session-strength beer of 3.5–4.5 % alcohol by volume will benefit from at least two weeks' **maturation**, and possibly four, but will remain fresh for two or three months as long as the barrel isn't tapped; a stronger beer should be matured for longer before it reaches its peak – six weeks, say, for a beer of 4.5–6% ABV – and will keep, untapped and properly stored, for up to six months. Impossibly strong beers will keep on improving almost indefinitely. Once tapped, though, no barrel should remain undrunk for more than three or four days. Once air gets into the headspace, the beer will start oxidising and will soon become undrinkable. If you can't get through 25 litres or five gallons of fresh cask beer in four days (and I'm not advocating drinking 10 pints a day, but you may be having a party), then it's probably better to bottle it (see page 83).

ASSESSING ALCOHOLIC STRENGTH

The alcoholic strength of all bottled alcoholic beverages now has to be declared on the label as percent alcohol by volume. This standardisation of strength declaration replaced years of confusion when the strength of wines and spirits used to be measured and declared as 'degrees proof' (or just 'proof' in North America), while the strength of beer was measured and declared as degrees of **original gravity**.

Actually, though, original gravity isn't a measure of alcoholic strength at all but of the sugar content, and therefore the *potential* strength, of the wort before fermentation. Potential strength doesn't quite equate to actual strength, because not all worts are fully attenuated – ie, not all the sugar in the wort is necessarily converted into alcohol. Some beers are more attenuated than others; some styles demand plenty of residual sugars, some none. So the original gravity will give you only a rough idea of the actual strength of your beer: an OG of 1048°, for instance, will correspond to 4.6–5% ABV.

Measuring the actual alcoholic content of the finished product is too tricky a business for the home brewer, which is why home brewers still have to use archaisms such as 'original gravity' and 'final gravity' when assessing the alcohol content of their beers.

Throughout the process of fermentation the gravity will sink as the sugars are consumed. This is a process you must monitor every day by using your **hydrometer**, basically a calibrated glass float that comes with a freestanding test-tube or trial jar, which you fill with a sample of your brew. The further the float sinks, the lower the gravity of the liquid in the jar. The **final gravity** will tell you, not the alcoholic strength of your beer, but how well attenuated it is (something that your palate will confirm). A good way of estimating alcoholic strength is simply to subtract the final gravity from the original or starting gravity and multiply by 0.13; so a beer that started at 1050° and stopped fermenting at 1010° will be (very roughly) 5.2% ABV.

Getting a correct reading from your hydrometer depends on the temperature of the sample measured, since temperature affects the density of the liquid. Most hydrometers are calibrated to operate at 20°C, so take the temperature of your sample before checking its gravity and, if it's too warm, stand the trial jar in a bowl of cold water for a few moments before checking the temperature again.

A hydrometer doesn't only measure the approximate strength of your beer; it also acts as a health-check: if the gravity stops falling you may have a stuck ferment (see page 39).

KEEPING RECORDS

You can have the best kit and the best ingredients, but even quite small variations in quantities, timings and temperatures can profoundly affect the quality of your beer. Keeping an accurate record of what you've added and when is the key to achieving every brewer's ambition, which is absolute consistency from brew to brew. That's why every brewer's most prized possession is not the kit, but the Brewer's Book.

Don't just record the timings and temperatures of your brew but of your brewery too: a recipe made in winter won't come out the same in summer. It's only by being meticulous in every detail that you can become that happiest of brewers who, when asked which is his or her favourite pint, can truthfully and more than a little smugly reply, 'My next one.'

BEER RECIPES

*Not just books, but whole libraries, could be and indeed
are taken up with beer recipes, and once you have caught
the bug you will inevitably start collecting them.*

Bput it would be over-ambitious here to do more than
provide a skeleton of basic recipes on which you can
improvise and extemporise to your heart's content. The
important thing is for you to familiarise yourself thoroughly
with the ingredients and methods laid out below so that,
when you are ready to fly, you'll know how to.

The recipes below are mostly concerned with quantities and
with requirements such as late-hopping. For details of the
processes, consult the previous chapter.

All mashes and boils are 90 minutes unless stated otherwise.
All the recipes given will make 25 litres or just over 5 gallons.

Dark Mild
Original gravity: 1035–36°. Final gravity: 1009°. ABV: 3.4%.
Pale malt: 3.5kg (7lb 8oz)
Crystal malt: 400g (14oz)
Black malt: 75g (3oz)
Hops: Fuggle 25g (1oz); Goldings 25g (1oz)

METHOD: Mash the malt in 10 litres (2¼ gallons) of treated
liquor for 90 minutes. Run off the wort into your boiler,
sparging the grains as you do so. Top up the boiler and boil
for 90 minutes, adding the hops as the wort comes to the boil.

Bitter

Original gravity: 1040°. Final gravity: 1009°. ABV: 4%.

Pale malt: 4.25kg (9lb)

Chocolate malt: 85g (3oz)

Hops: Challenger 15g (½oz); Target 25g (1oz)

Irish moss: 3g (1tsp)

METHOD: Mash with 10 litres (2¼ gallons) of treated liquor for 90 mins. Sparge, run off, and top up as above. Add the Challenger and half of the Target hops at the start of the boil; the remaining Target hops and Irish moss 10 minutes before the end.

Best Bitter

Original gravity: 1045°. Final gravity: 1008°. ABV: 4.8%.

Pale malt: 4.2kg (9lb)

Crystal malt: 290g (10oz)

Black malt: 30g (1oz)

White sugar: 450g (1lb)

Hops: Target 30g (1oz); Styrian Golding 15g (½oz)

Irish moss: 3g (1 tsp)

METHOD: Mash with 11 litres (2½ gallons) of treated liquor, sparge, run off, and top up as above.

Add the Target hops at the start of the boil and Irish moss 10 minutes before the end. Add the Styrian Golding hops while the wort is cooling after the boil.

Old Ale

Original gravity: 1055°. Final gravity: 1013°. ABV: 5.5%
Pale malt: 5kg (11lb)
Black malt: 320g (11oz)
Crystal malt: 175g (6oz)
Torrefied wheat: 90g (3oz)
White sugar: 290g (10oz)
Hops: Fuggle 70g (2½oz)

METHOD: Mash with 14 litres (3 gallons) of treated liquor; sparge, run off, and top up as above. Add 50g (1¾oz) of the Fuggle hops at the start of the boil and the remaining 20g (¾oz) of hops 10 minutes before the end.

Strong Ale

Original gravity: 1060°. Final gravity: 1015°. ABV: 5.8%.
Pale malt: 5.95kg (13lb)
Crystal malt: 325g (11½oz)
Chocolate malt: 230g (8oz)
Hops: Challenger 32g (1¼oz); Fuggle 15g (½oz);
    Goldings 15g (½oz)
Irish moss: 3g (1 tsp)

METHOD: Mash with 15 litres (3¼ gallons) of treated liquor; sparge, run off and top up as above. Add the Challenger and Fuggle hops at the start of the boil; add the Goldings hops and Irish moss 10 minutes before the end.

Strong Pale Ale

Original gravity: 1070°. Final gravity: 1015°. ABV: 7%.
Pale malt: 6.5kg (14lb)
Hops: East Kent Golding 400g (14oz)

METHOD: Mash in 15 litres (3¼ gallons) of treated liquor; sparge, run off, top up. Add the hops at the start of the boil and boil for 180 minutes.

Stout

Original gravity: 1050°. Final gravity: 1013°. ABV: 4.8%

Pale malt: 4.5kg (10lb)

Crystal malt: 450g (1lb))

Torrefied wheat: 435g (15oz)

Chocolate malt: 220g (7¾oz)

Hops: Fuggle 40g (1½oz); Progress 15g (½oz);
    Goldings 8g (¼oz)

METHOD: Mash in 15 litres (3¼ gallons) of treated liquor; sparge, run off and top up as above. Add the Fuggle hops at the start of the boil, the Progress hops 10 minutes before the end of the boil, and the Goldings after the boil.

Porter

Original gravity: 1055°. Final gravity: 1015°. ABV: 5.2%

Pale malt: 4.65kg (10lb)

Brown malt: 725g (1lb 9oz)

Crystal malt: 600g (1lb 5oz)

Chocolate malt: 120g (4¼oz)

Hops: Fuggle 95g (3½oz)

METHOD: Mash in 15 litres (3¼ gallons) of treated liquor; sparge, run off and top up as above. Add 70g (2½oz) of Fuggle hops at the start of the boil and the remaining 25g (1oz) 10 minutes before the end.

WINE: INGREDIENTS

Given sugar, water and a modicum of yeast nutrient, you can make wine out of pretty much anything. Prisoners (so I am told) ferment watered-down jam to make contraband hooch, while Tom's peapod burgundy was a running gag throughout the 1970s sitcom The Good Life.

In fact there's a snobbish assumption – going right back to Pliny the Elder in the first century AD, who in *De Rerum Natura* listed many kinds of 'artificial' wine including cider and perry – that the only 'real' wine, and certainly the only good wine, is made exclusively from grapes. This is simply not true. Fermented jam hooch made in jail might not stand up to a château-bottled claret; and using peapods that would serve better as animal fodder strikes me as a mark of desperation, but, provided you don't expect them to imitate grapes, the range of ingredients that will make not just good but excellent wines is almost endless.

THE WINEMAKER'S CALENDAR

The seasonality of home winemaking is one of its delights. Grapes can be harvested but once a year, but for the home winemaker nature offers a cornucopia of potential ingredients almost throughout the seasons. And not just the produce of your own garden, either: the footpaths, byways and even pavements around your home yield far more than blackberries. The winemaker who never goes for a walk without a couple of carrier bags can be pretty much guaranteed to come home with something to ferment, whatever the month.

Malting barley

Hops

Cox's orange pippin apples

Quince

Parsnips

Charles Ross apples

Conical fermenter

25-litre fermenter with tap

Wort chiller

Pressure barrel

Electric boiler

Stainless steel tank

Mash tun

Marrows

Rhubarb

Nettles

Five-litre glass demijohn
with basket

Bubble airlock

King keg
with bottom tap

Fruit press (iron)

Fruit press (wood)

Maxi fruit press

Bottle-filler
with three heads

Corking machine

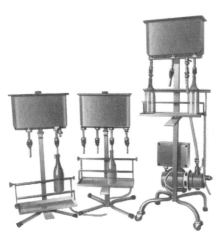

Linear bottle-fillers

Spring and early summer offer cherries, gooseberries, raspberries, strawberries, dandelions, elderflowers, hawthorn blossom, honeysuckle, nettles and thyme. In high summer there are marrows, apples, pears, apricots, damsons, gages, peaches, plums, bilberries, rose petals, elderberries, blackberries, and currants red, white and black. Autumn is the season of crab apples, quinces, hawthornberries, rosehips, rowanberries and sloes. Even winter has its bounty: carrots and parsnips will make strong, hearty, slow-maturing wines to lay down for next year or even the year after.

THE PROPERTIES OF DIFFERENT FRUITS

Base materials all have different properties, and while your own experiments will discover the best expression of their characters, here is a roll-call of the types of wines they usually create: **apricot**: dry white; **banana**: dessert white; **bilberry**: dry red; **blackberry**: dry red; **blackcurrant**: rich red; **carrot**: strong dry white; **cherry**: dry red; **crab apple**: dry white; **damson**: rich red; **elderberry**: dry red (a good keeper thanks to its tannin content); **greengage**: light dry white; **gooseberry**: dry white (picked green)/dessert white (picked ripe); **marrow**: full-bodied white; **nettle**: dry white; **parsnip**: dry sherry-ish white; **peach**: dessert white; **plum**: dry white, rosé, or red; **quince**: aromatic white; **raspberry**: dry rosé; **rhubarb**: light white, ideal for blending; **sloe**: rich red; **strawberry**: dry rosé.

The above list by no means exhausts your choice of base materials. **Apples** and **pears** can be used either to make wines by being chopped up and steeped in water and sugar, or to make true cider and perry by being pulped and pressed with no additions. **Citrus fruits**, even grapefruit, make more than

49

passable wines, as do **dried fruits**. And **cereals** – wheat, barley, even rye – can make wine as well as beer.

APPLES AND PEARS

To make true (all-juice) cider and perry you really need a scratter-mill and screw press (see page 15). This achieves a much better extraction than the standard home winemaker's method of chopping the fruit, macerating it in hot water and fermenting it with extra sugar. Without the additional sugar, though, the alcohol content will be lower – typically 6–7.5% ABV, but up to 8% ABV using dessert varieties in a good growing year (wet spring, hot summer).

Varieties are important. Ciders made from dessert apples are usually light-bodied and, while they can be delightfully aromatic, tend to lack body. This can be corrected by adding tea, which of course is a good source of the tannin that characterises West Country ciders. (The tannin adds to the cider's keeping qualities as well as to its body.) More satisfying, perhaps, is to locate a supply of crab apples or those little hard green wildings (domestic varieties gone wild) that you often find in country lanes. Added judiciously to your dessert apples, crabs or wildings will supply both tannin and acid to balance your blend. Of the dessert varieties, Bramleys make a light-bodied, fairly acid cider reminiscent at its best of an Italian soave, while Coxes provide a fuller body. Russets are also said to make good cider, particularly sparkling cider. Very sweet culinary varieties such as Gala, Golden Delicious and

Braeburn, though, don't: their high sugar content generates plenty of alcohol but very little flavour.

It is possible, however, to source genuine cider apple trees by mail order from a specialist nursery. There are hundreds of cider varieties, ranging in character from bittersweets relatively low in acid to bittersharps relatively high in it. Cidermakers generally blend half a dozen or more varieties, but there are varieties that will make a good cider on their own. Redstreak is one of the best, producing a gingery, lemony cider; Dabinett, Tremlett's Bitter and Yarlington Mill are also occasionally used to make single varietals; but the king of the cider apples is the Kingston Black which, almost uniquely, is also a good eater. For a list of specialist nurseries as well as more information on varieties, see www.ukcider.co.uk.

Perry is slightly more problematical. Dessert pears don't make good perry: because they're so soft when fully ripe they're difficult to press; they're very vulnerable to infection; and their juice has a tendency to blacken, which doesn't affect the flavour but is unappetising. Planting your own perry pear tree is not really an option since they can take 20 years or more to yield a worthwhile crop. Of the culinary pears, though, Robin (an East Anglian variety) makes a very acceptable perry.

GRAPES

The fact that we have argued the case of various fruits and flowers against the supremacy of the grape does not mean that we should turn our backs on grapes altogether. Grapes will grow in England, and English vineyards make some fabulous wines; properly looked after and grown against a south-facing wall, a vine or two should bear enough fruit to satisfy the needs of the home winemaker. You need to plant the right variety, however. Most garden centres only stock table grapes, which don't have the right character to make a decent wine (although, as they're high in sugar, they will make strong wines suitable for blending). Commercial vineyards in this country originally tended to select German varieties since our growing conditions were similar; but in recent years it has proved possible to grow grapes from further afield including French classics such as chardonnay and pinot noir. UK growers are also venturing into red wine production with some success, and English sparkling wines at their best rival champagne. Below is a selection of varieties favoured by commercial vineyards in England and Wales; see the English Wine Producers or UK Vineyards Association websites for nurseries that stock them.

White Varieties

AUXERROIS: Valued for its low acidity. Adds body to blended wines; can make sparkling wines or long-lasting table wines.

BACCHUS: A distinctive aromatic flavour and high sugar content. Grows well in the UK and makes a good single-varietal wine with New World sauvignon blanc characters.

CHARDONNAY: A fundamental ingredient of the finest sparkling wines, but also makes classic still wines.

HUXELREBE: A good cropper with good sugar levels. Susceptible to botrytis ('noble rot'), a fungus which dries out grapes, producing a higher sugar concentration and leading to very good dessert wines. High natural acidity and strong aromas of elderflowers, producing fruity wines that age well.

KERNER: Similar to riesling.

MADELEINE ANGEVINE: Designed for northern planting. Flowers late and is a reliable cropper. Useful for blending as it ages well and its low acidity complements higher acid varieties. On its own it produces light and fruity wines with a pronounced muscat bouquet.

MÜLLER-THURGAU (also known as **RIVANER**): The main grape in Liebfraumilch. Was among the first planted in the UK and was the single most widely grown variety for many years. Now less popular, being seen as producing unstylish wines. A vigorous early ripening variety, but can be a poor cropper.

ORION: One of a new generation of hybrid varieties bred both for wine quality and disease-resistance. Fruity and aromatic.

ORTEGA: Suits our climate but is prone to disease. Produces very full flavours and high natural sugars in wines that are rich and zesty with good balance. Good for blending with more neutral varieties.

PINOT BLANC: Ripens well and produces wine with full fruit flavours and crisp acidity. Crops heavily in most years. Can produce a style similar to chardonnay.

REICHSTEINER: Popular in the UK – the second most widely grown variety after seyval blanc. Ripens early, performs reliably, and is capable of producing large crops high in natural sugars. A little bland; often used for blending in both still and sparkling wines.

REGNER: Capable of good yields, ripens early with good sugars and relatively low acids – in short, an ideal candidate for our climate. Wine quality can be excellent.

SCHÖNBURGER: Very successful in the UK, producing wines with low acidity but high sugar and good muscat tones. When fully ripe it has a pink tinge. Its wines are distinctive, full-bodied and delicately flavoured.

SEYVAL BLANC: Crops heavily in this country even in cooler years, and has effective disease resistance. Good all rounder often used for blending, well suited to oak ageing, and used for still or sparkling wines. Single varietal wines offer crisp acidity with quite neutral flavours.

SIEGERREBE: Small-berried and intensely aromatic. Ripens sometimes to excessive levels and has a very dominating flavour. Often used to bolster blends.

Red Varieties

DORNFELDER: Notable for its colour and acidity. Grows well in the UK, showing that good red wine can be made here. Wines are usually fresh and fruity like syrah or gamay.

PINOT NOIR: One of the most ancient and noble of varieties. The classic grape for red Burgundy, but also an important element in UK sparkling wines.

REGENT: Another of the new hybrids bred for wine quality and disease resistance. A relatively new introduction to the UK whose wines produced have shown real promise, with low acidity, high sugar levels and good yields.

RONDO: Has adapted to UK conditions very well. Produces wines with very good colour and style and overtones of classic red varieties. Blends well with other varieties such as dornfelder and pinot noir and can be likened to a cross between tempranillo and syrah.

BLENDS AND ADDITIVES

Not all base materials possess rounded enough characters to stand on their own, rhubarb being the classic example. And just as cidermakers, the makers of Châteauneuf-du-Pape and the great cognac houses have become master blenders, so must you. Indeed, the process of experimenting with blends – blackberry and apple being one of the most common – is yet another of the joys of home winemaking.

As well as blending the base materials, the home winemaker must also use additives to correct their deficiencies. We are not talking about synthetic chemicals here. A judicious addition of lemon juice or zest will supply acidity where it is lacking, for instance, while raisins and sultanas are used to create a fuller body. And dessert apples typically make very light, dry wines which are all the better – and all the closer to true cider – for the addition of a cup of strong black tea.

HERBS AND SPICES

There's an old tradition of making wine entirely from herbs as a supposed tonic. Whether the medicinal properties of various herbs, even assuming they have any, are robust enough to survive fermentation I have no way of knowing. The drawback of making herb wines is the quantities involved. Three litres (5 pints) of lemon balm tips is an awful lot of lemon balm, but that's how much it takes to make 5 litres or a gallon of balm tip wine. Another drawback is that stronger-flavoured herbs don't actually make a very pleasant wine; their origins are medicinal and some of them taste that way. However, greatly reduced quantities of certain aromatic herbs do add a distinctive character to neutral wines. Star anise, bergamot, fennel, hops, lavender, verbena, borage and woodruff are all used – but there's an important caveat. If you add them early in the process you have no control over how their flavours will develop, and you can easily find that you've wasted a whole batch of wine. Better, perhaps, to make a strong tisane (infusion in hot water) of your chosen herb, draw off samples of your wine as it nears maturity and experiment, adding at first a little tisane to the wine sample, then a bit more until you get the flavour you prefer, and making a note of the proportions in order to add the same percentage by volume to the whole batch of wine. Add your preferred amount of tisane a few days before bottling and you'll end up with the subtle aroma and taste of your herb or herbs, rather than an overpowering smack in the mouth.

Spices – the Christmas spices especially – can also be used as late-added aromatics, preferably in sweet wines. Ginger, cloves, cinnamon and nutmeg are the classic mulling spices, and, added sparingly to damson wine, subtly evoke the season of goodwill whatever time of year you drink it. Caraway seeds, used similarly sparingly, will make a neutral white wine taste faintly of schnapps, while lightly cracked (*not* crushed) juniper berries, along with a little coriander and lemon balm, will recreate gin.

The best way to add spices is to make up a mulling-bag out of a square of muslin and suspend it in the must (the fermentable base for winemaking) on a piece of string. But never use powdered spices: they are insoluble and will form an unappealing dusty scum on the surface of your wine.

WINE: VINIFYING

Having chosen your ingredients, you are ready to turn them into wine. Well, to be accurate, you aren't – your yeast is.

YEAST

Yeast is a microfungus that is present pretty much everywhere. As soon as it finds a suitable sugary medium in which to make its home it will start reproducing rapidly, giving off alcohol and carbon dioxide as it does so. In nature, yeasts of many kinds thrive on rotting fruit. The brewers of Payottenland in Belgium still leave nature to get on with the business of fermentation; the downside of the natural approach is that alongside the desirable yeasts there are many undesirable ones, as well as other sundry microflora and fauna, that can produce some very unpleasant off-flavours.

Brewers, bakers and vintners in ancient times, though, learnt how to harvest benevolent yeasts and culture them. Today there are different strains of yeast that have been specialised to suit different types of wine, and your home brew shop will stock a wide selection either dried in individual sachets or liquid in vials, to be made up as directed. You can, if you have no shop nearby, use a teaspoon of dried baking yeast from a supermarket; or you might even choose to develop your own strain by skimming the barm from fermenting wine and keeping it in a sealed and sterilised jar in the fridge.

The main item in the yeast cell's diet is, of course, sugar. But yeast needs other nutrients, too, mainly nitrogen, phosphate, and vitamin B1. These occur naturally in many fruits, especially grapes, and fruit and vegetables grown in well-manured gardens. But most wild fruits, as well as most flowers and blossoms, are low in nutrients and may need a boost if a successful fermentation is to be achieved. This, incidentally, is where the old story about rats in cider comes from. Cidermakers of old somehow discovered that a piece of meat – usually bacon but quite possibly rat – thrown into the vat would restart a stuck ferment. They didn't know that it was the nitrogen given off during the decomposition of the meat that was doing the job by giving the yeast a good meal; they just knew that it worked. You, however, don't have to chop a packet of bacon into your demijohn: yeast nutrient tablets are available at your home brew shop.

WATER TREATMENT

Like beer, wine is mostly water. Grapes supply enough juice and don't need extra water; the same is true of apples and pears. But for all other recipes you will be resorting to the kitchen tap. Unlike the home brewer, though, the home winemaker doesn't really need to treat the water. The only caveat here is that, if you live in a hard water region, the level of dissolved chalk or calcium bicarbonate may increase the acidity of your wine beyond the optimum. If so, boil your water before use to precipitate the chalk and, when it's cool enough to handle, pour it carefully off the deposited sediment.

MACERATION

Washing the fruit thoroughly (enough to get rid of any mould spores) and either chopping or bruising it are the necessary first steps in preparing to release its fermentable and aromatic or flavouring components. Go back and re-read pages 14–16, 'Advanced Winemaking Equipment'.

As stated above, few fruits will make a satisfactory pure-juice wine and therefore then need to be macerated or steeped in water. Once the fruit has been steeped long enough, it is strained, and the resulting **must** is mixed with sufficient sugar to feed the yeast. The first stage of vinification therefore takes place in your bucket, where you introduce your processed fermentables to your (possibly) treated water. Different recipes call for the water either to be cold, warm or even boiling.

However there is one important adjustment to be made. Most fruits contain, in varying proportions, a sticky substance called **pectin** which, if left to itself, will inhibit a good extraction, cause excess frothing during fermentation, and create a haze in the finished wine. Pectin is what makes farm cider cloudy, and larger producers like to extract it from the juice to achieve a clear cider. There are three ways you can deal with pectin: either heat the fruit to 65°C before crushing to break down its cell walls; macerate it in warm water to allow its own pectin-destroying enzymes to get to work; or use a pectin enzyme from your home brew shop. Any of these procedures should mean you don't have to filter or add fining agents to your finished wine before bottling. Getting rid of pectin is particularly important where the recipe calls for the fruit either to be boiled or macerated in boiling water, since boiling destroys the fruit's natural pectin-digesting enzymes.

While the fruit pulp is macerating, **sugar** is added; with an eye to maintaining the correct quantities it should be remembered that 450g (1lb) of sugar adds 225ml (half a pint) to the volume of the must. Ordinary white granulated or caster sugar is ideal for most wines: pale brown sugars add colour but also have flavours of their own which you might not want. The sugar should be made into a syrup, using some of your must, before being added, otherwise it won't all dissolve. But do remember that, if you heat the syrup to make the sugar dissolve more quickly, it needs to be allowed to cool in a sealed, sterilised container before you add it. Some people add the whole quantity of sugar at this stage, but a better result, especially with sweet wines, will be achieved by adding only a third during maceration, followed by the second third as soon as the must has started fermenting, and the final third after the vigorous first or aerobic fermentation has finished.

Once maceration is complete, the must should be strained into the fermenter (a clean demijohn is ideal) to remove small floating particles. However, some recipes call for red wines to be 'fermented on the lees' – ie, left in the macerating bucket and the yeast added to the unstrained fruit pulp there – to get a better colour extraction from the skins of the fruit. It's also important to make more must than you need and reserve some in a sterilised sealed jar – a flip-top is ideal – to top up your wine as necessary. You can probably find an excellent culinary use for any leftovers.

TESTING AND ADJUSTING ACIDITY

Before starting to ferment your must, it's advisable to test it for acidity and make any adjustments necessary. A certain amount of acid is essential both to stimulate the fermentation and to balance the character of the finished product. Acid also protects against oxidation and bacterial infection and helps ensure clarity by precipitating proteins. The desirable acids are citric, malic and tartaric. Paradoxically, sweet wines need more acid and dry wines need less.

There are two ways of testing the acidity of your must. The simplest way is to use indicator paper, which comes with a graded chart to show you what pH value the change in colour of the paper represents. Must intended for the sweetest wines should have a pH value of just over 3, ranging to just under 4 for the driest. A more accurate way is to invest in a **titration kit**. Titration is a lot fiddlier, involving a test tube, an indicator and a reagent; but the kits come with instructions, and for many winemakers messing about with test tubes and chemicals is all part of the fun. The measure here is not pH but % tartaric: a desirable range for dry white wines is 0.55%T–0.65%T; for red wines 0.5%T–0.6%T; and for sweet wines 0.7%T–0.8%T. (Grape wines by and large will want to be very slightly more acidic.)

If your must isn't acidic enough, use **Acid Blend**, a ready-to-use packet of crystallised citric, malic and tartaric acids available from your home brew shop: a level teaspoonful per 5 litres (1 gallon) will increase the acidity by 0.15%T. If your must is too acid you can either dilute it slightly or use acid-reducing crystals – again, from your home brew shop.

As a general rule: flowers, root vegetables and rosehips are very low in natural acidity; elderberries and pears are low; apples, cherries, damsons, gages, peaches and plums are medium; blackberries, gooseberries, raspberries, strawberries and quince are high; and rhubarb, blackcurrants, redcurrants and whitecurrants are very high.

TANNIN

An important factor in how long your wine will take to mature, and how long it will keep after that, is its tannin content. Tannin is an astringent and mildly acidic polyphenol found on the surfaces of most plants, especially on immature leaves and unripe fruits. Its evolutionary function is to confer a bitter flavour that deters animals from eating the plant before it's ripe; and the bitterness can carry through to any beverages made from it. The bitterness of beer comes in part from the tannin in the hop, while the astringency of cider made from proper cider apples comes from their mouth-puckeringly high tannin content. Tannin also precipitates alkaloids and certain proteins, helping to prolong the life of drinks made from fermentable bases with high tannin content. Elderberries and red grape skins are high in tannin, which is why classic French red wines, especially clarets, age so well.

The tannin content supplied by nature is normally sufficient to do its job, but if tannin is low you can buy it in powdered or liquid form from your home brew shop, or simply add a cup of strong black tea to your must. It should be noted, though, that the tannin derived from tea is low in acid and, while it will add the right flavour to a cider made from dessert apples, it won't prolong its life.

ALCOHOLIC STRENGTH

The alcoholic strength of your finished wine is one of the hardest things to measure without a fully equipped laboratory, so instead you have to predict what it's going to be by careful control of the sugar content of your must. This is measured, as with beer, by using a hydrometer (see page 42), which will give you the specific gravity, or density, of the liquid.

Depending on what ingredients you are using, your must will already contain some sugar of its own, but not very much unless you are using grapes or making pure-juice cider or perry. So take a hydrometer reading before you add sugar, and then add only the amount you are going to need to finish up with the desired strength.

A rough guide to potential ABV (alcohol by volume) and the amount of sugar that a must of that gravity already contains is as follows:

1010°	0.9%	57g (2oz)
1020°	2.3%	200g (7oz)
1030°	3.7%	340g (12oz)
1040°	5.1%	480g (1lb 1oz)
1050°	6.5%	595g (1lb 5oz)
1060°	7.8%	708g (1lb 9oz)
1070°	9.2%	825g (1lb 13oz)
1080°	10.6%	930g (2lb 1oz)
1090°	12%	1.08kg (2lbs 6oz)
1100°	13.4%	1.19kg (2lb 10oz)
1110°	14.9%	1.3kg (2lb 14oz)

The ABV of a table wine can vary from 10% to 14%: much less than 10% and it won't keep; much more than 14% and the yeast will struggle to digest all the sugar. The starting gravity of the unfermented must should therefore be between 1075° and 1105°. A dry wine is going to require 2lb of sugar per gallon or 200g per litre; a medium wine requires 2lb 8oz per gallon or 280g per litre; a sweet wine requires 3lb per gallon or 290g per litre.

Remember, as we said before, that 450g (1lb) of sugar adds 225ml (half a pint) to the overall volume of must, so for 5 litres (a gallon) of dry wine you need 4 litres (7 pints) of must *plus* 900g (2lbs) sugar and so on. So to get a wine of, say, 12% ABV from a must of 1040°, you need to add 595g (1lb 5oz) of sugar to 4 litres (7 pints) of must.

FERMENTATION

Making a **starter** from a sachet of yeast is described on page 34; the winemaker should follow the same procedure as the brewer with one important difference. Ale wort is pitched with yeast at about 28°C and ferments briskly at a fairly high temperature. Wine likes to take its time and should be pitched at a lower temperature: 20°C is ideal.

The first or **aerobic fermentation** should be vigorous as the yeast multiplies rapidly and consumes the oxygen dissolved in the must. The working must should throw something of a head, although not as deep and dense as that of fermenting wort; so, if your recipe recommends that the must should be strained into a demijohn before it has started fermenting, remember to leave a good headspace, or the head will start overflowing messily. Aerobic fermentation can take anything from five days to a fortnight depending on the sugar content of the must, the vigour of the yeast strain and the ambient temperature (which should not be too hot: anything above 35°C will kill the yeast). If the aerobic fermentation is being carried out in an open fermenter, do remember to cover it securely either with a tight-fitting lid or a thick cloth loosely tied with string to protect it from fruit-flies. (If they manage to get at it, they will turn the whole batch to vinegar.)

Once it has slowed down, must that is fermenting on its lees (see page 60) should be strained into a clean demijohn; must that has had its first fermentation in a demijohn may either be transferred into a clean one or, if the aerobic stage has not been messy, simply topped up from your reserve of fresh must. If you're not racking it into a clean demijohn at this stage, wipe down the outside of the old demijohn with sterilising solution and make sure there is no sugary residue.

During the **anaerobic fermentation**, air with its population of unfriendly micro-organisms is the enemy: slowly fermenting wine is very vulnerable to infection and, unlike beer, it doesn't have the extra protection conferred by the acids contained in the hop. Two precautions must be taken to keep air at bay. The first is to fill the demijohn right up to the neck to leave as small an exposed surface area as possible. The second is to seal the neck with a tight-fitting rubber bung bored to take an airlock – a twisted piece of glass or plastic tubing with a U-bend filled with a few drops of sterilising solution to create a barrier impervious to microbes and fruit-flies.

The optimum temperature for anaerobic fermentation is around 15°C, and it should not be allowed to fluctuate by more than 2°C in either direction. At first, the yeast will work quite quickly, with a rapid stream of bubbles rising from the must and passing through the airlock. This is the period when the yeast is producing alcohol, and while it is doing so the wine should be left as far as possible undisturbed and in the dark. Depending on the amount of sugar in the must, the process can take several weeks or even months as the stream of bubbles gradually slows down, coming to a complete halt when all the sugar has gone. However, the yeast will not work far beyond 15–16% ABV, so if there is any unfermented sugar left at this stage you will end up with a strong, sweet wine.

NOTE: STUCK FERMENTATION

If the anaerobic fermentation appears to have ended before it should, you may have a 'stuck ferment'. The usual cause is temperature; but the must may also contain too little oxygen and/or too much CO_2, or the yeast may have run out of nutrient – a particularly common cause in wines made of wild fruit from unfertilised land. If you think your ferment has stuck, unbung the demijohn and give it a good stir to release any excess CO_2 and aerate the must slightly. Then add a fresh yeast starter with some yeast nutrient added, reseal it, and make sure the ambient temperature is right before leaving it to get going again. If your storage area is too warm, move it to somewhere cooler. If it's too cold, you can buy a heated mat for the demijohn to stand on.

CLEARING THE WINE

If the wine is slow to clear, there could be several causes. The most common is pectin (see page 59), which is easily dealt with by adding an enzyme bought from your home brew shop. You can test for pectin by drawing off a sample of about

30ml (1¼ fl oz) and mixing it with 100ml (4fl oz) of methylated spirits. This will cause the pectin to form a jelly-like clot. If no clot is formed, your yeast may not be settling properly, in which case the wine should be treated using **finings** from your home brew shop.

If the wine develops a hazy sheen, it may have been infected by lactic acid bacteria. This is a particular risk if any of the fruit was slightly rotten. In this case, treat the wine with three crushed Campden tablets per 5 litres (1 gallon); if they don't do the job, fine it as above.

If you see 'floaters' – long chains of bacterial cells – in the wine, it is turning to vinegar (acetifying) and there's not much you can do about it. If the infection is in its early stages, two crushed Campden tablets per 5 litres (1 gallon) of must might head it off. If that doesn't work, though, don't throw it out just yet: keep it well away from any other wine you may have on the go, sterilise the outside of the demijohn with your sterilising solution, and let it carry on working. The vinegar may be no good, in which case tip it away; on the other hand it may be excellent, in which case pretend it was intentional, bottle it in small fancy bottles, and give it to your friends and relatives at Christmas. The most common cause of acetification is contact with air; the lesson to be learnt is to fill the demijohns right up and bung them more tightly next time.

As well as lactic acid and vinegar-causing bacteria there is a host of other moulds, spores, bad yeasts and other micro-organisms that want to enjoy your wine as much as you do. There is always the risk that some will get through and that you will have to take appropriate action.

- **Mould** contamination is generally the result of not washing your fermentable base thoroughly enough. It also helps to confine your picking to dry, sunny days. Once you have mould, though, there's nothing you can do except throw the must away.
- **Film yeasts** are 'bad' yeasts which, as the name suggests, generate a film on the surface of the must and then break down the alcohol into CO_2 and water. They can be treated by adding two crushed Campden tablets per 5 litres (1 gallon) and, if they persist, by covering the surface of the must with a counter-film, as it were, of olive oil.
- **Oil or rope** looks worse than it is, affecting the appearance of the wine rather than its flavour. The viscous strands that form throughout the must can be dispelled by pouring it into a bucket and stirring it vigorously. After any sediment has precipitated, the must can be poured back into a clean demijohn.
- **Blackening or darkening** is caused by the use either of over-ripe or partly rotted fruit or of iron, copper or zinc utensils. The latter should simply be avoided. The former particularly affects perry and pear wine and can be prevented by adding two crushed Campden tablets to every 5 litres (1 gallon) of must.
- **Mouse** is a bacterial infection that creates a vile aftertaste. It makes the wine undrinkable, and nothing can be done about it except to tighten up your sterilisation and hygiene regimes.

WINE RECIPES

The recipes that follow are fairly standard and should always give perfectly acceptable results. But as you grow in experience and confidence you will undoubtedly want to tinker with them to produce wines of true individuality.

GRAPE WINE

Making wine from grapes is, in theory at least, the simplest process imaginable: all you have to do is press your grapes, collect the juice, add yeast and, once the vigorous first fermentation is over, rack the must into a clean demijohn. Nothing is ever that simple, though. You need an awful lot of grapes, for one thing: 6–7kg (12–15lbs) will only make 5 litres (1 gallon) or eight bottles of wine. And grapes grown in an English garden are likely to be a little too acid and not quite sweet enough to yield a first-rate wine: both problems will have to be assessed and addressed.

Before pressing, wash your grapes thoroughly, weed out any rotten, pulpy or discoloured berries, and remove all the stems and stalks (these will produce a very nasty bitterness). Once you have collected sufficient juice (remembering to set some aside in a sterilised sealed container in the fridge for topping-up purposes), dose it with a crushed Campden tablet and leave it to stand for 24 hours before fermentation.

At this stage, test the gravity of the must with a hydrometer. For a wine of about 12–13% ABV you will almost certainly need to add sugar: the following table will give you an idea of how much to add per 5 litres (1 gallon) of must.

1050°	425g (15oz)
1060°	310g (11oz)
1070°	200g (7oz)
1080°	113g (4oz)
1090°	57g (2oz)

If you have to use large quantities, make it up into a syrup using a little of the must and add it in three batches: one when you pitch the yeast, one a couple of days later, and the rest when the first aerobic fermentation is slowing down.

WHITE TABLE WINE

White grapes, of course, make white wine; but you can make white wine from black grapes as well by straining the juice off the pulp before fermentation. Must fermented without pulp, though, will be low in tannin once the first fermentation is over, and the wine is therefore best drunk young.

Once your pressed grapes complete with crushed Campden tablet have stood for 24 hours, draw off the must into a prepared demijohn. Then press the grape-pulp again to complete the extraction and add the juice to the demijohn. Leave enough headroom for the froth, and pitch with a yeast starter (see page 34). Cover the mouth of the demijohn with foil or clingfilm until froth starts to form, which should be within 48 hours. You can then safely uncover the mouth, as the CO_2 given off by the fermenting yeast will act as sufficient protection against bacteria. When the frothing stops, top up the demijohn with reserved must, seal it with a bung and airlock, wipe down the outside very thoroughly with sterilising solution, and transfer it to the airing cupboard.

Then leave it alone until the bubbles have stopped streaming through the airlock, which could be all winter. Once the bubbles have stopped, transfer the demijohn to a cold room – the garage, perhaps – for two weeks before siphoning the wine into a clean demijohn with airlock fitted, to clear and mature. Resist the temptation to taste it until spring arrives; if it's still too acid by then, take it back to the airing cupboard. After a while it should begin to ferment again, but only slightly: this is a malolactic fermentation, a secondary fermentation caused by friendly bacteria which convert malic acid to the less sour and metallic lactic acid, naturally de-acidifying and softening the wine. Once the bubbles stop, top up the demijohn with reserved must, seal it, and leave it to clear. Bottle it as soon as it's clear, and store the bottles on their sides until they're ready to drink, which should be within six months.

RED AND ROSÉ TABLE WINE

To make red or rosé wine, leave the must and pulp together in your fermenting bucket, tightly sealed (if the lid isn't airtight, stretch some clingfilm under it). Left for a couple of days, the skins will give up enough colour to make a delicate rosé; left for up to 10 days they will make a deep red. Do not leave for more than 10 days, or the wine may be bitter. The pulp must be completely covered by the must or the skins may dry out and acetify (turn acidic), spoiling the wine. One way of achieving this is to scoop the pulp into a muslin bag and keep the bag submerged in the must with a weight of some sort – *not* a metal one! Alternatively, press the bag of pulp down and give it a good swirl twice a day. Now follow the procedure for white wine, but use a brown-glass demijohn or cover the demijohn with a black binliner, because light will make the red colour fade. Red wine will have a greater tannin content than white; adjust the maturation period accordingly.

DESSERT WINE

Few treats can beat a good dessert wine, but it's quite tricky to make. The pH of the must needs to be slightly lower than for a dry wine – 3.1 rather than 3.3 – so test your must before fermentation and make any adjustment necessary.

Yeast is voracious stuff and will gorge itself on soluble sugar until, at between 16–18% ABV (depending on which type you buy), it literally eats itself to death. So to make a sweet wine you have to use enough sugar to kill the yeast but leave enough unfermented to sweeten the wine. This means your wine will be pretty strong stuff! The amount you are likely to need per 5 litres (1 gallon) is:

1050°	900g (32oz)
1060°	765g (27oz)
1070°	650g (23oz)
1080°	565g (20oz)
1090°	450g (16oz)
1100°	340g (12oz)

Additions should be made in very gradual stages: half of the sugar, dissolved in must, should be added in three stages as with table wine; and, whenever the fermentation seems to be slowing down, another 225g (8oz) can go in. When it has finished completely, taste it: if it's still not sweet enough you can add a little more dissolved sugar, but not too much at a time: you can always add more, but you can never add less!

A dessert wine needs plenty of body and flavour to balance all that alcohol, so the choice of grape variety is important: huxelrebe is ideal; auxerrois will do; and schönburger has pleasing muscat qualities but rather low acidity, which will need adjusting. If you're not growing any of these varieties you could add 250g (9oz) raisins to the grapes before pressing; alternatively, add honey to the wine after it has stopped fermenting but before laying it down to mature. If you do this, you need first to dissolve the honey in a little hot must or water, and mix up a few samples at different strengths before deciding how much to add to the wine. Because of its high alcohol content, dessert wine needs long maturation ... and a small glass!

CIDER, APPLE WINE AND PERRY

CIDER

Cider is easy to make: you simply pulp apples, press the pulp, transfer the juice to a fermenter, and let nature get on with it. Making *good* cider, though, takes a little more care.

First, select your apples. Probably these will be a mixture of cookers and eaters, whose juice will need to be manipulated somewhat to simulate true cider varieties. Modern dessert apples such as Braeburns aren't really suitable: they will provide plenty of alcohol, but far too much acid and not nearly enough tannin. Bramleys and Coxes, though, can easily be adjusted to create an ideal blend. Both are slightly more acid than cider varieties, but not quite as tannic. The acidity isn't a problem — aim for a pH value of 3.8–4 in your must. To increase the tannin content, add half a kilo (1lb) of crab apples or wildings per 7kg (15lbs) of cookers/eaters, or simply pour in a cup of very strong black tea.

You need about 6–7kg (12–15lbs) of apples to make 5 litres (1 gallon) of cider; but, as you need plenty of juice in reserve for topping up during fermentation, pick 8–10kg (16–20lbs). Leave them for a week after picking, during which time their tough cellulose fibre will soften slightly. Grate them in your scratter-mill, or pound them in a bucket, and press the resulting pulp or pomace in your screw press. When the pomace is pressed dry, rehydrate it with a little water and press it again to extract the last of the fermentable sugar.

Now dose your must with one crushed Campden tablet per 5 litres (1gallon) and leave it to stand for 24 hours in an airtight bucket or bin, sealed if necessary with clingfilm. This process will suppress the natural yeasts and bacteria in the juice, including bacteria that will produce vinegar and sundry other nasties. It's very important to keep air out, since the simple sugars contained in apples are irresistible to airborne microbes. Test the gravity of the juice: if it's below 1055°, add sugar dissolved in juice: 70g (2½oz) per gallon should raise the gravity by 5°, and how much you add depends on how strong you like your cider!

When the 24 hours is up, add yeast nutrient to ensure a good level of nitrogen – particularly important if your apples are from private gardens or hedgerows, which aren't as heavily fertilised as most orchards. To ensure a really clear cider (although many prefer it cloudy), add pectin-clearing enzyme as directed on the packet. Now pitch the juice with a suitable yeast – many people prefer champagne yeast for cider – and

transfer it straight away to a demijohn or any other vessel
to which an airlock can be fitted. Make sure the vessel is
brimming full: there should already be enough air dissolved in
the juice to get the aerobic ferment started. At this point,
cover the mouth of the fermenter with foil or film; within
48 hours aerobic fermentation should have started, and you
can uncover the mouth of the fermenter. During aerobic
fermentation, thick foam containing specks of solids will
emerge volcanically from the mouth of the fermenter; this is
normal, so don't be alarmed, but do keep wiping the outside
of the vessel with sterilised cloth. The eruption of foam will
protect the juice from infection during this phase; once it has
stopped, though, rack the juice into a clean fermenter, top it
right up from your reserve, and insert the bung and airlock.

Pick somewhere cool to store the fermenter throughout the
winter: cider ferments at about 15°C, but since fermentation
is an exothermic reaction (ie, it generates its own heat), an
ambient temperature of well below that won't be a problem.
Many commercial cidermakers conduct fermentation in
unheated barns or even outdoors. Check the gravity every few
days (always remembering to top up after taking a sample):
if the fermentation is rushed, the volatile aromatics will
evaporate and be carried off by the escaping CO_2; if it's too
slow, there's an opportunity for colonies of spoilage microbes
suppressed by the Campden tablets to regrow. Aim for the
gravity to fall by an average of one degree a day: if it's much
quicker than that, move the fermenter to somewhere cooler;
if slower, cover the fermenter with an old quilt or duvet.

Anaerobic fermentation can take two or three months
depending on the original gravity of the juice and temperature

of the surroundings. When it's finished, a third fermentation – a bacterial one – ought to begin. The malo-lactic fermentation converts malic acid into the much milder lactic acid. It's signified by the reappearance of a slow stream of bubbles through the airlock; if it doesn't happen you can actually buy the required bacterial culture from home brew shops.

Given the chance, your cider will ferment to complete dryness, which you probably won't want. There are two ways to prevent this. The simplest is to stop the fermentation at the point where you feel the cider is dry enough by adding one crushed Campden tablet per 5 litres (1 gallon). However the effect of the tablet can wear off and your cider can suddenly start working again, with potentially explosive consequences!

Better is to 'keeve' the cider. After milling, hold the pomace for 48 hours before pressing. The day after pressing, add 1ml (½ fl oz) of calcium chloride solution per 10 litres (2 gallons) of must. The must will then separate into three layers: sediment at the bottom, then clear juice, with pectin gel on top. This layer will have bound much of the nitrogen, thiamine and protein the yeast needs, so when you siphon off and pitch the clear juice it will ferment much more slowly.

Don't siphon it straight into a demijohn: transfer it instead into an open fermenter, because a thick brown cap will shortly form. Skim this off; if another forms, skim that off too. Once the must has stopped throwing these caps, rack it into a demijohn and apply the airlock as normal. Keep checking the gravity: by the time it has fallen to around 1025–30° the nutrient-deprived yeast should have given up the ghost and you will be left with a sweet but not terribly alcoholic cider.

APPLE WINE

If you don't have a proper mill and press for cider, apple wine can be made by the maceration or *dépense* method given for fruit wines (below) using 2.5–3kg (5–6½lbs) of roughly chopped fruit to 1.75–2kg (4–4½lbs) of sugar. Apples are full of pectin, so for a clear wine use a pectin enzyme. Yeast nutrient is also necessary unless your apples come from a commercial orchard. Apple wine is an excellent carrier for spices, especially the Christmas spices – ginger, cinnamon, and nutmeg. The apples can also be mixed with pears for a slightly sweeter, richer wine.

PERRY

Perry is made in the same way as cider, with one important difference: pears are by and large much more heavily laden with tannin than apples. To precipitate the excess, the juice should be stood on the pulp overnight in a cool or even a cold place before being racked off into the fermenter; and it should undergo its secondary fermentation at as low a temperature as can be achieved. It will then throw a more or less continuous sediment, from which it has be racked off frequently.

Other problems derive from the pear's deficiency in soluble nitrogen and from its high acidity, both of which can result in a slow fermentation and consequent spoilage by opportunistic microbes. So make sure the acidity is at the correct level – a pH value of 3.9–4 – and use plenty of yeast nutrient. A good perry is well worth the effort, though: in particular, pears contain sorbitol, an unfermentable sugar that gives the finished drink a rich, almost buttery mouthfeel.

You can mix apples and pears to make a drink waggishly named 'pyder'. In fact a good proportion of pears in your cider will supply the tannin missing in dessert apples, and the sorbitol content will compensate for the yeast's propensity towards complete dryness; although the must of such a blend will need its acidity checking and reducing.

FRUIT, VEGETABLE AND FLOWER WINES

For most home winemakers, using various fruits – either grown in your own garden or collected wild – is often the main attraction. But one thing you should remember is that these fruits probably don't come from trees that have been as well-manured or fertilised as they would be in a commercial orchard, so they will be low in nitrogen. It's therefore almost always essential to use a yeast nutrient.

STONE FRUIT

Cherries, plums, damsons, apricots, peaches, gages and even sloes make a great variety of wines (see page 49). Their common characteristic is that they're rich in pectin, and the use of boiling water in the steeping process destroys what natural enzymes they have; so they all require the use of a pectin enzyme to achieve clarity. They also need pressing with care so that the stones remain unbroken.

The method is much the same for all of them. They should be put in muslin bags and gently bruised with a rolling pin to break the skins and pressed either in a screw press or, if you don't have one, by wringing. The pulp should then be steeped in a gallon of boiling water in a bucket or bin and left, well covered and with a crushed Campden tablet added, for 2–3 days to get the maximum extraction. Finally, dissolved sugar, pectin enzyme, yeast nutrient and yeast starter are added.

For white wine, strain the must into a demijohn for the first fermentation. When the frothing subsides, transfer to a clean demijohn and top up; insert bung with airlock, and ferment. For red wine, leave the must on the lees (in the macerating bucket) for the first fermentation, and siphon into a demijohn with bung and airlock for the secondary fermentation.

The proportions of fruit to sugar per gallon are as follows:

Apricot: 1.8kg (4lbs) fruit to 1.25kg (2lbs 13oz) sugar.

Cherry: 3.5kg (8lbs) fruit to 1.5kg (3lbs 8oz) sugar (morello cherries are best if you can get them).

Damson: 1.8kg (4lbs) fruit to 1.5kg (3lbs) sugar.

Greengage: 1.8kg (4lbs) fruit to 1.8kg (4lbs) sugar.

Peach: 1.8kg (4lbs) fruit to 1.35kg (3lbs) sugar (add the zest and juice of a lemon and 225g (8oz) minced raisins to the must).

Plum: as greengage.

Sloe: 1.35kg (3lbs) fruit to 1.5kg (3lbs 8oz) sugar (add 225g (8oz) minced raisins to the must).

SOFT FRUIT

Blackberries, raspberries and strawberries all make excellent light rosés. If you don't grow your own raspberries and don't fancy paying shop prices for them, add a couple of punnets to blackberries. Tannin-rich elderberries make the claret of the home winemaking world. Gooseberries make light, dry, slightly acidic wines like a good hock. Red- and whitecurrants make delightful light summer wines, while blackcurrants, as you'd expect, are gutsier. And they're all so easy to process – they don't even need pressing. Only elderberries, which have to be de-stemmed, and strawberries, which should have their green caps cut off, need any special treatment. Otherwise, just wrap them in muslin and roll them lightly, or simply mash them up in your bucket with a potato masher, then treat as stone fruit above. All are high in pectin, so add the enzyme as well as yeast nutrient.

The proportions of fruit to sugar are as follows:

Blackberry: 1.8kg (4lbs) fruit to 1.8kg (4lbs) sugar.

Blackcurrant: 1.35kg (3lbs) fruit to 1.35kg (3lbs) sugar.

Elderberry: 900g (2lbs) fruit to 1.35kg (3lbs) sugar; juice and zest of two oranges.

Gooseberry (green): 1.35kg (3lbs) fruit to 1.35kg (3lbs) sugar.

Gooseberry (pink): 1.8kg (4lbs) fruit to 1.35kg (3lbs) sugar.

Raspberry: 1.8kg (4lbs) fruit to 1.35kg (3lbs) sugar.

Redcurrant/whitecurrant: 1.8kg (4lbs) fruit to 1.1kg (2lbs 8oz) sugar.

Strawberry: 1.8kg (4lbs) fruit to 1.35kg (3lbs) sugar; zest and juice of a lemon.

VEGETABLES AND RHUBARB

Vegetable wines are often the butt of jokes, but root vegetables are packed with sugar and flavour and make very good wines. They do, however, tend to lack acidity – a fault that can be corrected with the zest and juice of two oranges and two lemons per gallon of must. Rhubarb in particular makes an excellent base for flavoured and spiced wines.

Vegetable wines are less likely to need yeast nutrient, since they tend to come from well-fertilised plots rich in the essential nitrogen. However, they can be difficult to clear, especially thanks to the boiling involved, and do need a pectin enzyme.

The method is to chop, slice or grate them without peeling; boil them with about 60g (2oz) bruised root ginger until just soft but not mushy; let them cool, and then add the zest and juice, pectin enzyme, and dissolved sugar. Add a crushed Campden tablet, allow the must to stand well-covered for a day, pitch with the yeast starter and strain into a demijohn with bung and airlock. When the first fermentation is finished, rack into a clean demijohn and top up. Marrows, however, do not need to be boiled but merely need to be covered with boiling water and allowed to stand.

The proportions of fruit to sugar are as follows.
Carrot: 1.8kg (4lbs) to 1.8kg (4lbs) sugar (ideally demerara).
Marrow: 2.3kg (5lbs) to 1.35kg (3lbs) sugar.
Parsnip: 1.35kg (3lbs) to 1.35kg (3lbs) sugar.
Potato: 2.3kg (5lbs) to 1.8kg (4lbs) sugar (ideally demerara).
Rhubarb: 2.3kg (5lbs) to 1.6kg (3lbs 8oz) sugar (ideally demerara).

FLOWERS

Flowers make surprisingly good wines; but of course they contain no fermentable materials themselves: this has to be supplied in the form of 1.1kg (2½lbs) white sugar and 225g

(8oz) minced sultanas or 110g (4oz) white wine concentrate per gallon. The juice and zest of two lemons is also required for acidity, and, while yeast nutrient is essential, pectin enzyme is surplus to requirements. Flower wines are best drunk fresh and should be matured for no more then 3–4 months before bottling.

The method is the same for all recipes. Wash the flowers well and measure them not by weight but loosely packed in a pint jug or mug. Cover them with 5 litres (1 gallon) of boiling water, cool, and add a crushed Campden tablet. Draw off a litre (1¾ pints) of the must and heat it to boiling to dissolve your sugar. Cool the syrup to blood heat (37°C) and return it to the must; add the minced sultanas or white wine concentrate, the lemon zest and juice, and yeast nutrient. Pitch with a yeast starter, cover well, and leave until the first fermentation has died down. Siphon off into a demijohn, insert bung and airlock, and leave to ferment.

The quantities of flower heads (all stems removed) to be used are as follows:
Carnation: 1.75 litres (3 pints)
Dandelion: 2.25 litres (4 pints)
Elderflower: 600ml (1 pint)
Hawthorn blossom: 2.25 litres (4 pints)
Honeysuckle: 1.75 litres (3 pints)
Marigold: 2.25 litres (4 pints)
Primrose: 3.4 litres (6 pints)
Rose petal: 2.25 litres (4 pints)

THE FINISHED PRODUCT

'Ripeness is all,' said the bard, and it's true – both beer and
wine are all the better for maturing to full ripeness.

Not to over-ripeness, mind: some wines deserve to be laid down, while others need to be drunk fresh; some beers will improve with age, whereas others will go stale. But, old or new, your beers and wines will spoil if they're not bottled carefully. Include the date of laying down on your label, as well as the type of wine or beer.

HYGIENE

Hygiene is critical at every stage, but never more so than when transferring the liquid from one container to another. This is when the beer or wine is most at risk of coming into contact with the air and its hostile microflora and fauna; and this is the stage when it will be passing through pipes and taps and into bottles, all of which are difficult to clean effectively. Soak all pipes, tubes, bottles, taps, etc. in sterilising solution for at least 20 minutes before use, and remember to have to hand your bottle brush and a stiff toothbrush! Any cloths you use for wiping down surfaces should be boiled beforehand.

BOTTLE CHOICE

I'm assuming that you will have no difficulty in amassing a stock of ordinary wine bottles – it takes eight to bottle the contents of a 5-litre (1-gallon) demijohn. For still wines, any old bottle will do, although it adds a certain flourish to use the correct ones: round-shouldered green glass Bordeaux-style for gutsy tannic reds; slope-shouldered Burgundy-style for lighter-bodied reds; tall, slim white glass Alsace-style for delicate, aromatic whites and so on. (Red wine will lose its colour if put in white glass bottles.) Quite apart from the sheer style, using different bottles for different kinds of wine helps you to organise your cellar more efficiently!

But ordinary wine bottles will only do for still wines and ciders. Once gas pressure is introduced into the equation, you need more robust bottles that won't explode in the cellar. Ideally, these should be proper champagne bottles, because your sparkling wines won't be, as most commercial sparkling wines are, merely a stabilised still wine with a measured amount of CO_2 artificially injected. Yours will actually still be fermenting slightly, which means that the gas pressure will be slowly increasing over months and even years. An ordinary asti or cava bottle probably won't explode – but only probably. Your home brew shop will stock strong enough bottles: they're fairly expensive, but can be re-used many times.

Beer bottles of almost any sort will be strong enough, but they really should be brown. Beer suffers from a condition called 'lightstruck' if exposed to UV light: it damages the flavour, and only brown glass is able to filter UV effectively. Green glass will do at a pinch, if your beer is stored in a place that is dark.

CLOSURES

Corks are still, for most people, the best closure for still wines. Soaked overnight before use, they swell to the point where they fit the neck of the bottle perfectly and, provided the bottle is stored on its side to keep the cork damp, they will exclude air almost indefinitely. A tip: soak your corks in a jar with a lid filled right to the brim – corks float, and if you just leave them in a bowl of water the upper part won't be saturated enough to swell. Also, boil the soaking water beforehand and sterilise the corks before use.

If using traditional corks, you need help to drive them home safely and securely. You can't just hit them directly with a camping mallet (never use a hammer) because you risk breaking the lip of the bottle. The simplest device is a cork-flogger – basically a wooden cup which holds the cork securely and protects the bottle while you wield your mallet (two or three taps not a single mighty blow!). Freestanding lever-operated corking machines are quicker and safer and come in a range of sizes and prices.

There are alternatives. Home brew supply shops sell the sort of corks used in port and sherry bottles, which can be reused; but they don't fit as tightly as a proper cork, which means (a) that the bottle can't be stored on its side, and (b) there is a consequent risk of air getting in and spoiling the wine. The same applies to plastic stoppers. Reusable corks and plastic stoppers are fine for wines intended to be drunk young, but if you are using them you should always fit a heat-shrunk capsule as well, just to be sure they're airtight. These capsules are simplicity themselves to fit: you just slip one over the neck of the bottle and rotate it briefly in the jet of steam from a boiling kettle.

For sparkling wines, a proper champagne mushroom with wire cage cannot be beaten. Not only does it look the part – it has stood the test of time in practical terms. A proper champagne bottle closed with a proper cork and cage can withstand incredible pressure almost indefinitely. Home brew shops stock both the corks and cages, and also foil capsules, which have no practical advantage but set the bottle off nicely. Screw tops, especially the thin metal screw tops used in modern wine and spirits bottles, are by and large unsuitable. They can balloon under pressure and eventually fly off; if anything's gone wrong with your bottling, even a supposedly still wine might have enough puff left in it to do damage. Cidermakers often use a heavy-duty plastic screw top which is strong enough but not especially attractive.

For beer bottles, traditional crimped metal crown corks – or crown caps, as they are now called since the internal cork pad has been replaced by plastic – were invented for the job and are the ideal closure. All home brew shops supply them; the only drawback is that you need a crown capper to apply them. This can cost from around £10 for a simple hand-held version up to £70 for a freestanding lever-operated model – which will, however, also drive corks into wine bottles quickly and easily. A good alternative is to build up a collection of swing-topped bottles, for which you can buy replacement washers as they wear out.

BOTTLING STILL WINES

How still is your still wine? Residual fermentation may not be apparent to the naked eye when checking the airlock for bubbles; and as you will be laying your bottled wine down for months or possibly years there will be plenty of time for even a slight fermentation to build up to a dangerous level of pressure. So, a few days before you plan to bottle, move your demijohn(s) to a warm room and see if there are signs of activity starting up again. If any telltale bubbles show themselves in the airlock, either dose the wine with one crushed Campden tablet per 5 litres (1 gallon) or move it back to the airing cupboard and wait a little while longer.

The usual method of bottling wine is simply to siphon it straight from demijohn to bottle. Siphoning, though, is a messy business, and carries a slight additional risk of infection. Better, in my view, to strain the wine carefully into a vessel with a tap, give it a couple of days to settle, and then pour it straight into the bottles via the tap and a good old-fashioned funnel. Not only is this quicker and less messy, it has the added advantage of giving the wine one last polish. But do remember to keep the tap clean!

BOTTLING SPARKLING WINES

Apples, pears, rhubarb and gooseberries make excellent sparkling wines; and of course elderflower champagne simply *has* to have bubbles. The simplest way of getting that added fizz is to bottle the wine before it has finished fermenting; but, given that the pressure will continue to increase as the fermentation goes on, this is really only suitable for wines you intend to drink straight away, especially elderflower. Even with these, the continuing fermentation makes opening the bottles very tricky: more and more CO_2 dissolves in the wine, and as soon as the cap is loosened it releases itself in a violent and spectacular fountain.

I prefer to let the wine ferment to dryness, rack it into a fresh demijohn, leave it to mature for a few weeks, and then bottle it (in champagne bottles, of course) with a level teaspoon of sugar and a small yeast starter. This will work just enough to produce the fizz without the fountain.

You could, of course, go the whole hog and bottle your sparklers by the *méthode champenoise*. This involves fermenting the wine out, bottling it, and then storing it upside down so that the sediment forms in the neck. You then freeze the neck so that when you open the bottle all the sediment comes out in one lump, make up the volume with a starter as above, and reclose the bottle with a mushroom cork and wire cage. It all seems a bit of a faff, somehow.

STORING DRAUGHT BEER AND CIDER

Pouring a professional-looking pint of real ale – clear as a bell, and with a good head on it – is one of the more satisfying moments in home brewing. But few home brewers would take matters to the extremes of screwing a handpump to the dining table and connecting it to a proper oak or steel firkin (9-gallon cask) with yards of pipework. Most are happy to settle for a plastic pressure barrel. This, basically, is a 25-litre (5 gallon) plastic container. It sits upright, not on its side, and has a tap (spigot) at the bottom and a screw-top lid with a built-in outlet valve to vent any surplus CO_2 given off during secondary fermentation. Thoroughly sterilised, filled right up, lid screwed on firmly and stored in a cool, dark place, a pressure barrel will protect even a low-alcohol beer – a mild, say – for two or three months, and a strong ale virtually forever. The strongest ale will keep on improving like a good claret.

Once broached, though, the barrel's contents will start oxidising fairly quickly and need to be drunk within three or four days. If the prospect of getting through five gallons of a 6% ABV beer in four days seems a bit daunting, you can either bottle it (see below) or buy a device from your home brew shop that flushes the headspace with protective gas – either CO_2 or a CO_2-nitrogen mixture.

The polycask is almost identical to the pressure barrel except that it has no venting valve. Polycasks are widely used by cidermakers and, unbroached, the contents will keep for two or three years or even longer, depending on their alcoholic strength, acidity, and tannin content. Even broached, a polycask of strong West Country cider can still be perfectly drinkable after three months.

A polypin is another good option for keeping draught beer and cider. A collapsible plastic box with a tap, it excludes air almost perfectly – exactly like the collapsible bag in a supermarket winebox. The drawback is that polypins were designed as one-trip vessels because, being cubes, they have eight hard-to-reach corners where bacteria and moulds just love to set up home. The best advice is only to use one as a short-term storage option; help protect the contents by adding a crushed Campden tablet; and leave a litre of sterilising solution inside when the polypin is not in use.

BOTTLING BEER

Beer shouldn't really be bottled straight from the fermenter: it may well have enough condition (ongoing fermentation) to blow its cork out unless it is properly matured and, if necessary, fined. Beer to be bottled should be racked first into a barrel and left for anything from two weeks for a session-strength bitter to two months or more for a strong stout, porter, or old ale.

Once the beer has matured, vent it by releasing the cap and leave it for a day. Then draw off a sample to check for clarity and condition. It should have settled out to brightness by now; if not, fine it using isinglass or a vegetarian alternative; see page 41 for the amounts and method.

If no head appears in the glass when drawing off the sample, add a little priming sugar just to create some condition (make it continue fermenting a little longer). The best sugar to prime with is a solution of 50g (2oz) of malt extract made up to 225ml (half a pint) with boiling water and left to cool. Ordinary white caster sugar will do instead of malt extract; and darker sugars and even molasses will add a certain something to dark and strong beers.

To bottle the beer, use a siphon tube attached to your tap, long enough to reach the bottom of your bottles. This further reduces the risk of infection. Cap the bottles straight away and store them not, as with wine, on their side, but upright. The sediment will then sink to the bottom of the bottle and will not be disturbed by careful pouring.

Leave session-strength beers for a month to come into condition before drinking. The stronger the beer, the longer it will take to mature. Very strong beers will continue to improve, like very tannic wines, for years: I have sampled Thomas Hardy Ale at 20 years and found it stunning!

GLOSSARY

airlock: tube containing sterilising solution inserted through bung of fermenter. Excludes bacteria.

alcohol: intoxicating carbohydrate produced by fermentation.

alpha acid: antibacterial component of hop cones.

aromatics: flavour components of hops, fruit, flowers, etc.

attenuation: the degree to which malt sugar is converted to alcohol during fermentation.

barley: the cereal most commonly used to make malt.

barrel: brewers' standard unit of measurement – 1.63hl (36 gallons).

boiler: vessel in which wort is boiled with hops.

bottle-conditioned: beer bottled with live yeast; continues a slow fermentation in the bottle.

bottom fermentation: fermentation by lager yeast. Typically cooler than ale yeast.

brewery-conditioned ('keg'): beer matured at the brewery then filtered, pasteurised, and artificially carbonated.

bright beer: cask-conditioned beer that is re-racked, typically into a polypin, shortly before serving, usually on boats or trains or at parties, fêtes, etc.

Burtonisation: adjustment of mineral content of brewing liquor, usually to remove chalk and add gypsum.

Campden tablets: sodium metabisulphite in tablet form. Antibacterial; also inhibit yeast growth.

cask-conditioned: ale that is racked from fermenter to cask with living yeast in order to continue a slow secondary fermentation. Also called 'real ale'.

chalk: calcium carbonate. Component of hard water that damages brewing equipment, raises acidity of wort, and inhibits fermentation. Precipitated by boiling.

concentrate: fruit juice reduced to a thick gel by boiling, then rehydrated before fermentation. Can produce unwanted caramel flavour, especially in cider.

condition(ing): the secondary fermentation or maturing period of beer, and the dissolution of CO_2 in the beer.

cork flogger: wooden cap placed over wine cork during bottling to prevent damage to bottle.

copper: see Boiler.

demijohn: 5-litre (1-gallon) glass or plastic jar used for fermenting and maturing wine.

dry-hopping: adding a handful of late hops to cask beer to intensify its aroma.

extract: concentrated malt.

extraction: solution of fermentable sugars in wort or must by mashing or macerating, and of hop-bittering agents and aromatics by boiling.

final gravity: see Gravity.

finings: substances used to clear microparticles and protein haze from beer or wine.

firkin: cask of 9-gallon capacity.

gravity: measure of the concentration of fermentable (typically sugars) and non-fermentable materials in any liquid. Original gravity (OG) is the concentration of solids before fermentation. Final gravity is the concentration of solids after fermentation. The difference multiplied by 0.13 is a good indicator of the final product's ABV.

grist: blend of malt used in mash or hops used in boil.

gypsum: calcium sulphate. Reduces acidity of wort and aids fermentation and clarification.

hops: climbing plant related to cannabis and nettles.

Irish moss: a red alga whose mucilaginous body attracts microparticles. Used late in the boil to help clarify the beer.

isinglass: collagen derived from the swim-bladder of fish, used as finings. Its positive ions attract and precipitate negatively charged yeast, hop and protein particles.

keg: see Brewery-conditioned.

kettle: see Boiler.

kilderkin: a cask of 18-gallon capacity.

liquor: brewing water.

malt: grains of cereal, usually barley, germinated by steeping in hot water and then dried in a kiln. The process converts the insoluble starch in the grain into fermentable maltose sugar, and the flavour and character of the malt is determined by the temperature and duration of the drying.

mash: ground malt steeped in hot water.

must: the fermentable base for winemaking, either pure fruit juice or more commonly the liquid produced by macerating fruit or vegetables with sugar.

Milton Sterilising Fluid: proprietary liquid used to sterilise babies' bottles and in brewing used as an antibacterial agent.

OG: see Gravity.

pasteurisation: the process of halting fermentation and killing bacteria, mould spores, etc. using heat. The process can ruin delicate drinks, especially cider, and provided you have sterilised your bottling equipment and bottles effectively is not necessary.

pectin: a mucilaginous gel present in many fruits, especially apples. If untreated, it will cause cloudiness.

perry: cider made with pears instead of apples.

pitch: to add yeast to the fermentable base (must or wort).

polycask: plastic barrel-shaped container, usually five gallons, with a screw cap on the top and a tap near the bottom.

polypin: an airtight collapsible box with a tap, commonly used for dispensing bright beer.

pomace: milled apples or pears ready for pressing.

pressure barrel: basically a polycask with a vent in the cap to allow surplus CO_2 to escape.

priming: the process of adding a little yeast and sugar to casks or bottles to prolong secondary fermentation and add sparkle to the beer (or wine).

racking: the process of transferring beer or cider from the fermenter to the cask.

scratter-mill: a hopper containing rotating blades turned by a hand crank or small motor, used for milling apples, pears and root vegetables.

sodium metabisulphite: antibacterial agent preferred by most home brewers.

sparging: the process of sprinkling hot water over the grist while running out of the mash tun. Improves extraction.

stock ale: a strong, well-matured ale usually used for blending with other beers.

tannin: an astringent, slightly acidic polyphenol found on the surfaces of most plants. The acids help to protect wine and cider against microbial contamination while ageing. Red wines fermented on the pulped skins contain more tannin than white wines and thus age better. Cider apples have a far higher concentration of tannin than dessert apples.

top fermentation: fermentation using ale yeast, which works more quickly and at a higher temperature than lager yeast and often throws a large, rocky head.

wort: the fermentable base of beer.

yeast: a fungal micro-organism that digests sugar to produce CO_2 and ethanol.

yeast nutrient: the nitrogen, phosphate and vitamin B1 needed by yeast to reproduce.

INDEX